SADDLEBACK
PUBLISHING • INC.

UP-TO-SPE[ED]

NUMBER SENSE

by Jennifer G. Bove
and Thomas H. Hatch

UP-TO-SPEED MATH

Number Sense
Operations
Working with Fractions, Decimals, and Percents
Geometry and Measurement
Patterns, Functions, and Algebra
Data Analysis and Reasoning

created by **Kent Publishing Services, Inc.**
designed by **Signature Design Group, Inc.**

SADDLEBACK PUBLISHING, INC.
Three Watson
Irvine, CA 92618-2767

E-mail: info@sdlback.com
Website: www.sdlback.com

ISBN 1-56254-383-0

Printed in the United States of America

TABLE OF CONTENTS

LESSON 1 UNIT 1

POWERS OF TEN

George has to clear snow from the parking space in front of his house. The space is a square 10 feet wide and 10 feet long.

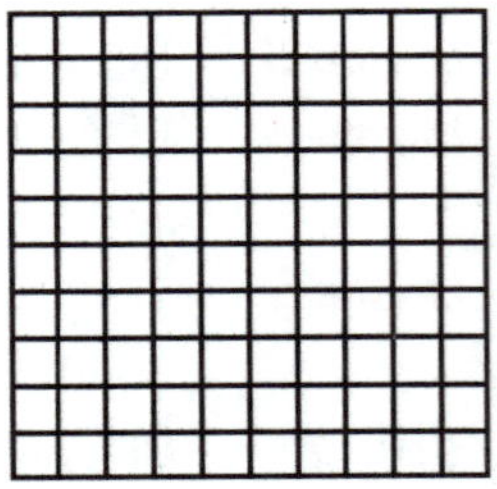

How many square feet are there in George's parking space?

$10 \times 10 = 100$ square feet

We can also use exponents to show this relationship.

$10^2 = 100$ square feet

An exponent tells how many times a number is used as a factor.

$$10 = 10 \times 1 = 10^1$$
$$100 = 10 \times 10 = 10^2$$
$$1000 = 10 \times 10 \times 10 = 10^3$$
$$10{,}000 = 10 \times 10 \times 10 \times 10 = 10^4$$

A Write a number in exponent form for each picture.

1. ______________

2. ______________

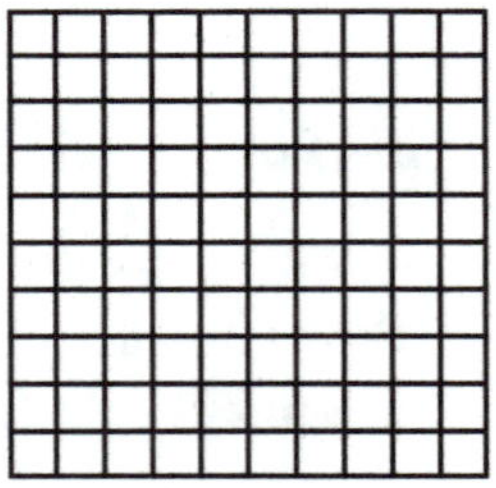

3. ______________

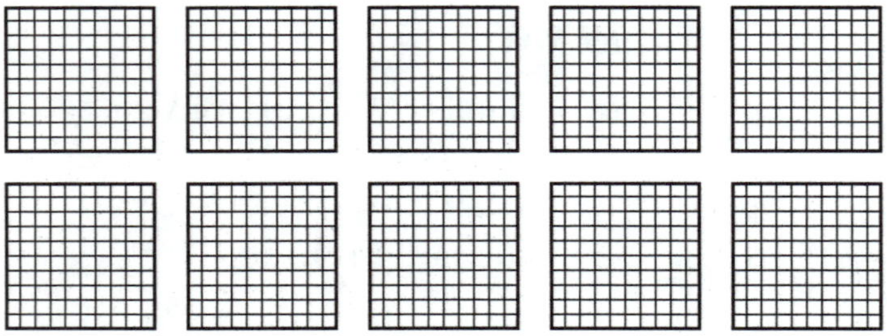

B Complete the information in the chart.

		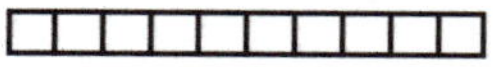	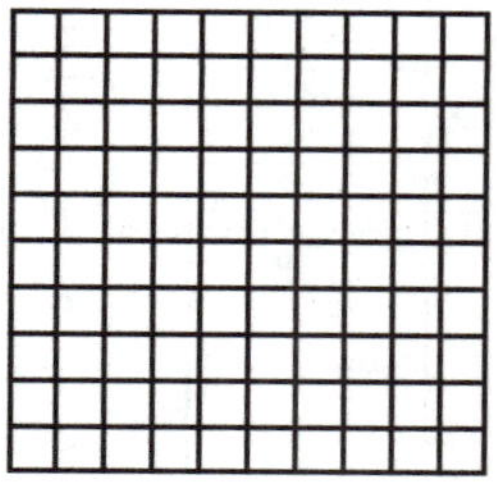	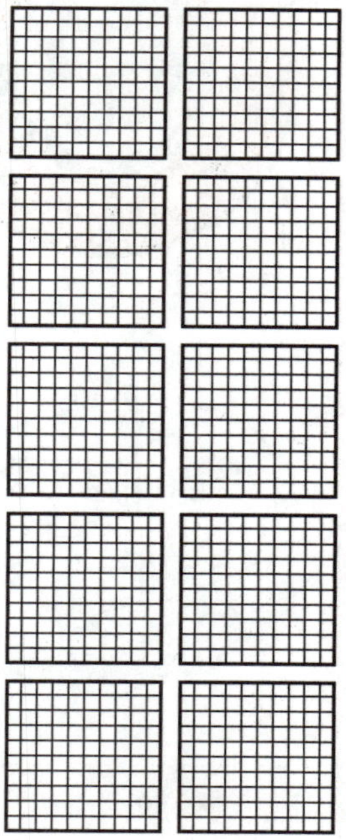
number	1	1.	100	1000
factored form	1 x 1	10 x 1	10 x 10	2.
exponent form	10^0	10^1	3.	10^3
power of 10	zero	first	second "ten squared"	third "ten cubed"

C Use the chart above to answer these questions.

1. How many tens in 10 squared? _______
2. How many hundreds in 1000? _______
3. How many ones in 10^0? ___________
4. Which power of 10 equals 100? ______
5. Which power of 10 equals 10? _______
6. If 10^2 is 100 and 10^3 is 1000, what is 10^4? _______________
7. What is 10^5? _______________
8. What is 10^6? _______________

LESSON 2 UNIT 1

PLACE VALUE TO THOUSANDS

The new Vanstar weighs 3,286 pounds. But it can pull a loaded horse trailer that weighs 5,470 pounds. Each digit in these numbers has a value that depends on where it appears.

	thousands	hundreds	tens	ones	Standard Form	Expanded Form
	10^3	10^2	10^1	10^0		
Weight	3	2	8	6	3,286	3,000 + 200 + 80 + 6
Towing Capacity	5	4	7	0	5,470	5,000 + 400 + 70 + 0

The 8 in the number for weight is in the tens column. It means:

8 tens 8×10 8×10^1 80

The 5 in the number for towing capacity is in the thousands column. It means:

5 thousands $5 \times 1{,}000$ 5×10^3 5,000

A Look at the chart above. Complete these statements.

1. The 3 in the number for weight means:

   3 thousands   ______ × 1,000 $3 \times 10^{\square}$ ______

2. The 2 in the number for weight means:

   2 ______   ______ × 100 $2 \times 10^{\square}$ 200

3. The 4 in the number for towing capacity means:

   4 hundreds   4 × ______ $4 \times 10^{\square}$ 400

B Write the place value for the underlined digit.

1. 8, 2̲46 — 2 Hundreds
2. 7̲53 — ______
3. 5, 793̲ — ______
4. 9̲, 301 — ______

Write in standard form.

5. 500 + 30 + 7 ______
6. 4,000 + 40 + 6 ______
7. seven thousand two hundred nine ______
8. four thousand four hundred seventy one ______

Write in expanded form.

9. 439 ______
10. 7,302 ______

C Explain what each underlined digit means. The first two have been done for you.

1. 4̲,346	4 x 1,000	4×10^3
2. 368̲	8 ones	8×10^0
3. 19̲2	______	______
4. 5̲, 655	______	______
5. 8,76̲5	______	______
6. 3̲1	______	______

LESSON 3 UNIT 1

PLACE VALUE TO MILLIONS

Each digit in a number has a value that depends on where it appears in the number.

	The United States of America						
	millions	hundred thousands	ten thousands	thousands	hundreds	tens	ones
	10^6	10^5	10^4	10^3	10^2	10^1	10^0
Number of States Admitted to the Union						5	0
Total Area (in sq. miles)	3	7	1	7	5	2	2

The 5 in the number of states is in the tens column. It means:

5 tens   5 x 10   5 x 10^1 50

The 3 in the number for total area is in the millions column. It means:

3 millions   3 x 1,000,000   3 x 10^6 3,000,000

A Look at the chart. Complete the statements on the next page.

Delaware is known as the "First State" because it was the first to ratify the United States Constitution. Delaware was admitted to the Union on December 7, 1787. Here are some facts about Delaware:

	The State of Delaware					
	hundred thousands	ten thousands	thousands	hundreds	tens	ones
	10^5	10^4	10^3	10^2	10^1	10^0
Population - 731,581	7	3	1	5	8	1
Land Area - 1,955 sq. mi.			1	9	5	5
Students Attending Public School - 110,549	1	1	0	5	4	9

1. The 7 in the population number means —

7 hundred thousands ______ × 100,000 $7 \times 10^{\square}$ ______

2. The 9 in the land area number means —

9 ______ ______ × 100 $9 \times 10^{\square}$ ______

3. The 9 in the number of public school students means —

9 ______ 9 × ______ $9 \times 10^{\square}$ ______

B Explain what each underlined digit means in two different ways. The first two have been done for you.

1. 5<u>5</u>4,346 — 5 x 10,000 — 5×10^4
2. 36,<u>8</u>90 — 8 hundred — 8×10^2
3. 1<u>9</u>2 ______ ______
4. <u>6</u>5, 655 ______ ______
5. <u>1</u>,908,765 ______ ______

Here are three ways to write the same number:

standard form	14,263
expanded form	10,000 + 4,000 + 200 + 60 + 3
expanded form using exponents	$(1 \times 10^4) + (4 \times 10^3) + (2 \times 10^2) + (6 \times 10) + (3 \times 1)$

C Write these numbers in standard form.

1. $(3 \times 10^4) + (5 \times 10^3) + (3 \times 10^2) + (6 \times 10) + (7 \times 1)$ ______
2. $(1 \times 10^5) + (4 \times 10^4) + (2 \times 10^3) + (7 \times 10^2) + (6 \times 10) + (1 \times 1)$ ______
3. $(2 \times 10^3) + (7 \times 10^2) + (3 \times 10) + (1 \times 1)$ ______
4. $(9 \times 10^4) + (9 \times 10^3) + (0 \times 10^2) + (0 \times 10) + (6 \times 1)$ ______
5. $(1 \times 10^3) + (4 \times 10^2) + (9 \times 10) + (2 \times 1)$ ______
6. $(3 \times 10^2) + (6 \times 10) + (0 \times 1)$ ______

LESSON 4 UNIT 1

PLACE VALUE IN DECIMALS AND MONEY

Here's the bill Lettie received when she helped her son buy his first car. Look at the place value chart. Places to the right of the decimal point stand for values between 0 and 1.

Hudson Motors
Guaranteed Used Cars

1993 Pickup	$2,500.00
sales tax @ 6.75%	168.75
Title and License fees	76.35
total	$2,745.10

	thousands	hundreds	tens	ones		tenths	hundredths
	10^2	10^2	10^1	10^0	.	10^{-1}	10^{-2}
sales tax percent - 6.75%				6	.	7	5
fees -$76.35			7	6	.	3	5
sales tax cost - $168.75		1	6	8	.	7	5
cost of truck - $2,500.00	2	5	0	0	.	0	0
total - $2745.10	2	7	4	5	.	1	0

A Complete the sentences using the chart above. Circle your answers.

1. The 7 in the cost of fees means—

   A 70 dollars B 70 cents C 7 dollars D 7 cents

2. The 5 in the cost of the truck means—

   A 50 dollars B 50 cents C 500 dollars D 5000 dollars

3. The 1 in the total cost means—

   A 10 dollars B 1 cent C 1 dollar D 10 cents

4. The 5 in the sales tax cost means—

   A 50 cents B 5 dollars C 5 tenths of a dollar D 5 hundreths of a dollar

5. The 5 in the sales tax percent means—

A .5% B 5% C 0.05% D 50%

6. The 6 in the sales tax percent means—

A 0.6% B 6 dollars C 6% D 6 cents

7. The 8 in the sales tax cost means—

A 80 dollars B 8 dollars C 8 cents D 8 percent

8. The 2 in the cost of the truck means—

A $2000 B $200 C $20 D $2

B Complete the sentences. Refer to the chart below if you need help.

ten thousands	thousands	hundreds	tens	ones		tenths	hundredths	thousandths	ten thousandths
10^4	10^2	10^2	10^1	10^0	.	10^{-1}	10^{-2}	10^{-2}	10^{-4}

1. Which decimal number shows 5 tenths and 15 thousandths?

A 0.0515 B 0.515 C 50.0015 D 5.015

2. Which decimal has a 7 in the hundredths place?

A 0.077 B 780.74 C 0.707 D 70.767

3. Which decimal includes a digit that stands for 4×10^4?

A 4.6452 B 0.4855 C 40,223.04 D 431.0974

4. Which decimal number represents 3 tens, 3 tenths, and 3 thousandths?

A 0.30303 B 3.0303 C 30.303 D 33.03

5. Which set shows the decimals in order from greatest to least?

A {0.05, 5.0, 0.5, 1.5} C {5.0, 1.5, 0.5, 0.05}

B {5.0, 1.5, 0.05, 0.5} D {0.05, 0.5, 1.5, 5.0}

Lesson 5 Unit 1

TEST TRACK

A Chose the best answer for each question below.

1. In which number does the 3 have the greatest value?

   Ⓐ 137 Ⓑ 1.302 Ⓒ 23.5 Ⓓ $\frac{1}{3}$

2. Write in expanded form using exponents: 53,405

   Ⓐ $(5 \times 10^3) + (3 \times 10^2) + (4 \times 10) + (5 \times 1)$

   Ⓑ $(5 \times 10^4) + (3 \times 10^3) + (4 \times 10^2) + (5 \times 1)$

   Ⓒ $(5^3 \times 10^3) + (40 \times 10^2) + (5 \times 1)$

   Ⓓ $(5 \times 10^4) + (5 \times 10^2) + (4 \times 10) + (5 \times 1)$

3. In which number does the 4 have the least value?

   Ⓐ 1,504 Ⓑ $\frac{4}{5}$ Ⓒ 5.94 Ⓓ 5.954

4. Which number includes a 6 that stands for 6×10^2?

   Ⓐ 5,066,004 Ⓑ 5,605,650 Ⓒ 4,405,566 Ⓓ 6,544,054

5. Evaluate the expression $(1 \times 10^4) + (3 \times 10^3) + (2 \times 10^2) + 5 =$

   Ⓐ 13,205 Ⓑ 13,250 Ⓒ 10,325 Ⓓ 10,370

6. Which expression represents 11,456 written in expanded form?

   Ⓐ $(1 \times 10^4) + (1 \times 10^3) + (4 \times 10^2) + (5 \times 10^1) + (6 \times 1)$

   Ⓑ $(11 \times 10^4) + 400 + 50 + 6$

   Ⓒ $(1 \times 10^3) + (1 \times 10^2) + (4 \times 10^1) + (5 \times 10) + (6 \times 1)$

   Ⓓ $(11 \times 10^3) + (4 \times 10^2) + 5 + 6$

7. Which number contains a 5 that stands for 5×10^5?

Ⓐ 505,650 Ⓑ 656, 500 Ⓒ 5,060,005 Ⓓ 650

8. Which number does NOT contain a 7 that stands for 7×10^2?

Ⓐ 17,707 Ⓑ 74,750 Ⓒ 77,731 Ⓓ 77,077

9. Which decimal number shows 67 tenths?

Ⓐ 0.067 Ⓑ 6.67 Ⓒ 0.607 Ⓓ 6.7

10. Which decimal number represents 4 ones, 5 tenths, and 6 thousandths?

Ⓐ 0.40506 Ⓑ 0.456 Ⓒ 4.506 Ⓓ 45.06

11. Which set shows the decimals in order from least to greatest?

Ⓐ {0.03, 3.0, 0.3, 3.5}

Ⓑ {3.5, 3.0, 0.03, 0.3}

Ⓒ {3.0, 3.5, 0.3, 0.03}

Ⓓ {0.03, 0.3, 3.0, 3.5}

12. 25 cents is the same as:

Ⓐ 2.5 dollars Ⓒ 25 tenths of a dollar

Ⓑ 25 hundredths of a dollar Ⓓ 25 dollars

B Write in expanded form using exponents.

1. 153,604 ______________________________

2. 3,096,659 ______________________________

LESSON 1 UNIT 2

MULTIPLES

When you multiply two or more numbers, their product is a *multiple* of those numbers.

2 x 4 = 8 **8 is a multiple of 2 and a multiple of 4**

MULTIPLES OF 2				
2	x	1	=	2
2	x	2	=	4
2	x	3	=	6
2	x	4	=	8
2	x	5	=	10
2	x	6	=	12
2	x	7	=	14
2	x	8	=	16
2	x	9	=	18
2	x	10	=	20

The products in the last column are all multiples of 2.

The first ten multiples of 2 are:
2, 4, 6, 8, 10, 12, 14, 16, 18, 20

If you continue the times table, every product will be a multiple of 2.

A Find the first ten multiples for each number.

1. **3**

3 x 1 = ☐	3 x 6 = ☐
3 x 2 = ☐	3 x 7 = ☐
3 x 3 = ☐	3 x 8 = ☐
3 x 4 = ☐	3 x 9 = ☐
3 x 5 = ☐	3 x 10 = ☐

2. **7**

B Find three multiples for each number.

1.	4	4 x 5 = ☐	4 x 7 = ☐	4 x 3 = ☐
2.	5	5 x 1 = ☐	5 x 6 = ☐	5 x 11 = ☐
3.	9	9 x 4 = ☐	9 x 10 = ☐	9 x 5 = ☐

C Find the first three multiples for each number.

1. 6   6 ______ ______
2. 10 ______ ______ ______
3. 8 ______ ______ ______
4. 11 ______ ______ ______
5. 12 ______ ______ ______
6. 13 ______ ______ ______

Sometimes you need to find a common multiple for two or more numbers. Find two common multiples of 2 and 5.

2 x 1 = 2	2 x 6 = 12	5 x 1 = 5
2 x 2 = 4	2 x 7 = 14	**5 x 2 = 10**
2 x 3 = 6	2 x 8 = 16	5 x 3 = 15
2 x 4 = 8	2 x 9 = 18	**5 x 4 = 20**
2 x 5 = 10	**2 x 10 = 20**	

Two common multiples of 2 and 5 are 10 and 20.

D Find common multiples.

1. A common multiple of 3 and 5 is ______
2. Two common multiples of 5 and 6 are ______
3. Three common multiples of 2, 3, and 5 are ______

HINT

For any set of numbers, you can find one common multiple by multiplying them all together.

2 x 3 x 4 = 24

24 is a common multiple of 2, 3, and 4

LESSON 2 UNIT 2

DIVISIBILITY RULES

Here's a set of rules that can help you figure out if one number is evenly divisible by another.

$$12 \div 3 = 4$$

dividend (12) — divisor (3) — quotient (4)

If your divisor is:	Then, it will divide evenly into:
2	all even numbers.
3	all numbers whose digits add up to a multiple of 3. Example: 126: 1+2+6=9. 9 is a multiple of 3, so 126 is divisible by 3
4	all numbers whose last two digits form a number that is a multiple of 4 Example: 316. 16 is a multiple of 4, so 316 is divisible by 4.
5	all whole numbers with a 0 or a 5 in the ones place. Example: 65, 125, and 250 are divisible by 5.
6	all numbers that are divisible by both 2 and 3. Example: 432 is even, so it is divisible by 2. And, 4+3+2=9, so it is divisible by 3 as well. Therefore, 432 is divisible by 6.
9	all numbers whose digits add up to a multiple of 9. Example: 2763. 2+7+6+3=18. 18 is a multiple of 9, so 2763 is divisible by 9.
10	all whole numbers with a 0 in the ones place. Example: 60, 110, and 2280 are divisible by 10.

A Choose the answer that best completes each statement.

1. If a number is divisible by 6, it is also divisible by —

   A 2 B 3 C 2 and 3 D 2 and 4

2. Numbers that are divisible by 10 are also divisible by —

   A 5 B 2 and 5 C 4 D 2, 4, and 5

3. Numbers that are divisible by 9 are also divisible by —

   A 6 B 4 C 2 D 3

4. 633 is divisible by —

   A 2, 3, and 6 B 3 and 6 C 6 D 3

B List the answers to each question.

1. Of the first ten multiples of 6, which are also divisible by 4?

   ____

2. Of the first ten multiples of 3, which are also divisible by 9?

   ____

3. Of the first ten counting numbers, which can be evenly divided into 160?

   ____

4. Of the first ten counting numbers, which can be evenly divided into 165?

   ____

C Solve. Show your work. Write the divisiblity rule that applies to each problem.

1. 2232 ÷ 2 = ______________________________
2. 2232 ÷ 3 = ______________________________
3. 2232 ÷ 4 = ______________________________
4. 2232 ÷ 6 = ______________________________
5. 2232 ÷ 9 = ______________________________

LESSON 3 UNIT 2

FACTORS

Numbers multiplied by each other are called *factors* of the product. The product is multiple of each of the factors.

factor of 6		factor of 6		multiple of 2 and 3
2	x	3	=	6

Every number has itself and 1 as factors. To find other factors, use division. Factors divide evenly into their multiples with no remainder.

Find the factors of 30.

Step One: 30 and 1 are factors.

Step Two: Divide by the counting numbers, starting with 2.

15
2)30
20
10
10
0

There is no remainder.

2 x 15 = 30

2 and 15 are factors of 30

10
3)30
30
0

There is no remainder.

3 x 10 = 30

3 and 10 are factors of 30.

7
4)30
28
2

There is a remainder.

4 is not a factor of 30.

6
5)30
30
0

There is no remainder.

5 x 6 = 30

5 and 6 are factors of 30.

4
7)30
28
2

There is a remainder.

7 is not a factor of 30.

3
8)30
24
6

There is a remainder.

8 is not a factor of 30.

3
9)30
27
3

There is a remainder.

9 is not a factor of 30.

The factors of 30 are:
1, 2, 3, 5, 6, 10, 15, 30

A Find the factors.

1. Find the factors of 20.

   Step One: 1, 20

   Step Two: Divide by counting numbers, starting with 2.

   Factors of 20: ______________

2. Find the factors of 15.

   Step One: 1, 15

   Step Two: Divide to find other factors.

   Factors of 15: ______________

3. Find the factors of 18.

   Step One: 1, 18

   Step Two: Divide to find other factors.

   Factors of 18: ______________

B To find common factors, find the factors of each number then compare to see which factors are the same.

1. What are the common factors of 15 and 20?

   Factors of 15: ____________ Factors of 20: ____________

   Common factors of 15 and 20: ____________

2. What are the common factors of 24 and 40?

   Factors of 24: __________________ Factors of 40: __________________

   Common factors of 24 and 40: __________________

LESSON 4 UNIT 2

PRIME FACTORS

Every number has itself and 1 as factors. A *prime number* has only itself and 1 as factors.

Factors of 5:	5, 1	5 has only two factors: itself and 1. 5 is a prime number.
Factors of 9:	9, 1, 3	9 has three factors: itself, 1, and 3. 9 is not a prime number.

Numbers that are not prime are called **composite numbers**.

You can tell if a number is prime or composite when you look for factors. If you find more than two factors, the number is composite. If you can only find two factors (itself and 1), the number is prime.

Study this list of the prime numbers under 20.

2, 3, 5, 7, 11, 13, 17, 19

A Test each number to see if it is prime or composite.

1. **33** Factors ______________________

   33 is a ______________ number.

2. **133** Factors ______________________

   133 is a ______________ number.

3. **46** Factors ______________________

   46 is a ______________ number.

4. **23** Factors ______________________

   23 is a ______________ number.

HINT

Remember to try division when you are looking for factors. Divide by the counting numbers, starting with 2. If you divide without a remainder, you have found a factor.

Every composite number is the product of prime numbers. To show all the prime factors of a composite number, make a factor tree. What are the prime factors of 36?

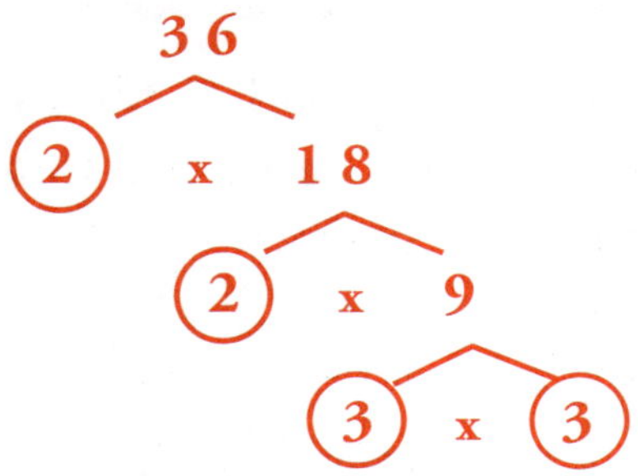

The prime factorization of 36 is 2 x 2 x 3 x 3.

When you write prime factorizations, use exponents whenever a prime factor is used more than once. The exponent tells how many times the prime factor is used. The prime factorization of 36 is $2^2 \times 3^2$.

B Complete each factor tree, then write the prime factorization with and without exponents.

1. 20

   2 x

   2 x

   prime factorization of 20 ____________

   using exponents ____________________

2. 48

   2 x

   2 x

   2 x

   2 x

   prime factorization of 48 ____________

   using exponents ____________________

3. 56

   x 28

   x 14

   x 7

   prime factorization of 56 ____________

   using exponents ____________________

4. 81

   3 x

   3 x

   3 x

   prime factorization of 81 ____________

   using exponents ____________________

LESSON 5 UNIT 2

LCM & GCF

A pizza shop wants 2, 3, or 4 people to be able to share a pizza equally. They do not want the slices to be too small. Into how many slices should they cut their pizza?

To Solve: Find the Least Common Multiple of 2, 3, and 4.

Using the times tables, find the multiples of each number. The least common multiple you find will be the LCM.

Multiples of 2: 2, 4, 6, 8, 10, **12**, 14, 16, 18....
Multiples of 3: 3, 6, 9, **12**....
Multiples of 4: 4, 8, **12**....
LCM of 2, 3, and 4: 12

Cut the pizza into 12 pieces.

A Find the LCM for each set of numbers.

1. 3, 7

   Multiples of 3: ________________

   Multiples of 7 ________________

   The LCM of 3 and 7 is __________

2. 2, 4, 6

   Multiples of 2: ________________

   Multiples of 4: ________________

   Multiples of 6: ________________

   The LCM of 2, 4, and 6 is ________

3. 3, 5

   Multiples of 3: ________________

   Multiples of 5: ________________

   The LCM of 3 and 5 is __________

HINT

If you start with the smallest number, you don't have to list as many multiples for the greater numbers.

You can also find the Greatest Common Factor for a set of numbers. What is the Greatest Common Factor for 12, 24, and 36?

List all the factors of each number to find the common factors. Then see which common factor is the greatest.

Factors of 12:	1, 2, 3, 4, 6, 12
Factors of 24:	1, 2, 3, 4, 6, 8, 12, 24
Factors of 36:	1, 2, 3, 4, 6, 9, 12, 18, 36
Common Factors:	1, 2, 3, 4, 6,12

Greatest Common Factor (GCF): 12

B Find the GCF for each set of numbers.

1. 5, 10, 15

Factors of 5: ________________

Factors of 10: ________________

Factors of 15: ________________

Common Factors: ________________

GCF: ________________

2. 8, 20

Factors of 8: ________________

Factors of 20: ________________

Common Factors: ________________

GCF: ________________

3. 16, 24, 40

Factors of 16: ________________

Factors of 24: ________________

Factors of 40: ________________

Common Factors: ________________

GCF: ________________

4. 9, 30, 39

Factors of 9: ________________

Factors of 30: ________________

Factors of 39: ________________

Common Factors: ________________

GCF: ________________

5. 26, 28, 48

Factors of 26: ________________

Factors of 28: ________________

Factors of 48: ________________

Common Factors: ________________

GCF: ________________

6. 42, 63, 84

Factors of 42: ________________

Factors of 63: ________________

Factors of 84: ________________

Common Factors: ________________

GCF: ________________

LESSON 6 UNIT 2

TEST TRACK

A Find the least common multiple for each set of numbers. Show your work.

WORK SPACE

1. 3, 4 ____________
2. 2, 5, 6 ____________
3. 5, 8 ____________
4. 3, 5, 9 ____________
5. 3, 7, 9 ____________

B Find the greatest common factor for each set of numbers. Show your work.

WORK SPACE

1. 52, 39 ____________
2. 72, 96 ____________
3. 26, 104 ____________
4. 34, 68, 170 ____________
5. 24, 32, 40 ____________

C Write the factors for each number.

1. 32 ____________
2. 37 ____________
3. 65 ____________
4. 66 ____________
5. 81 ____________
6. 91 ____________

D Write *prime* or *composite*.

1. 73 ____________
2. 55 ____________
3. 90 ____________
4. 2 ____________
5. 15 ____________
6. 23 ____________

E Create a factor tree for each number. Write the prime factorization with and without exponents.

1. 24

____

____

2. 96

____

____

3. 48

____

____

4. 144

____

____

LESSON 1 UNIT 3

NAMING FRACTIONS

Three friends want to share a brownie for dessert. They will have to cut it into three equal parts. Each part will be called *one-third* of the *whole* brownie. There are three thirds in one whole.

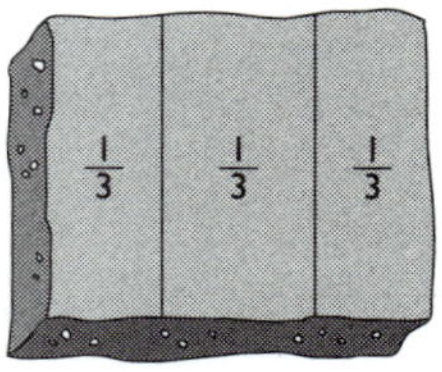

one third is written $\frac{1}{3}$ and means **one out of three**.

$\frac{1}{3}$ is a fraction. Every fraction has two parts:

1 — numerator (the number of parts we are working with)
3 — denominator (the total number of parts in the whole or group)

Here are some other ways of showing $\frac{1}{3}$:

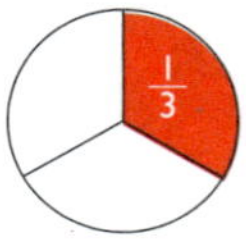

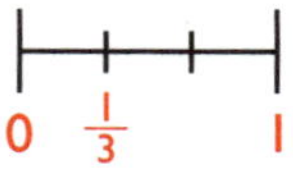

A

Here are some common fractions. Fill in the missing information. Count the shaded parts to write the numerator (top number). Count all the parts to write the denominator (bottom number).

1. 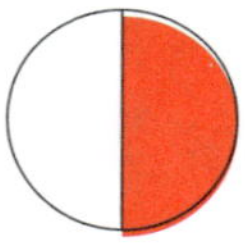 $\frac{1}{2}$ ______ one-half ______ one out of two ______

2. 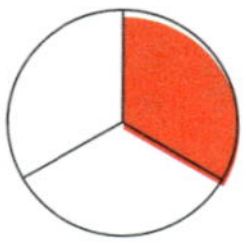 $\frac{1}{3}$ ______ one out of three

3. 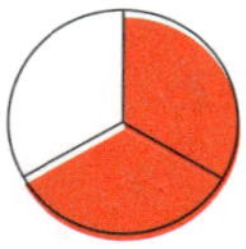 ______ two-thirds two out of three

4. 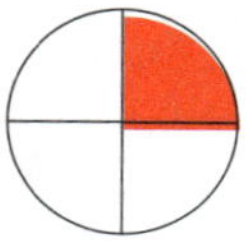 ______ one-fourth ______

5.		$\frac{3}{4}$	______	______
6.	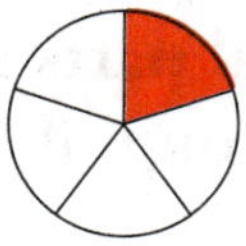	______	one-fifth	______
7.	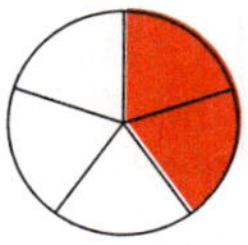	$\frac{2}{5}$	______	______

B Fractions can be written for *parts of a whole* or *members of a group*. Write a fraction for the shaded members of each group.

1.	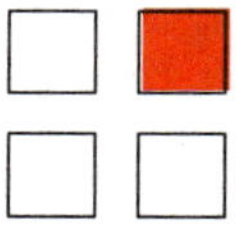	$\frac{1}{4}$	one-fourth	one out of four
2.	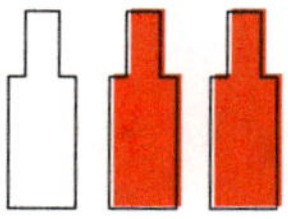	______	______	______
3.	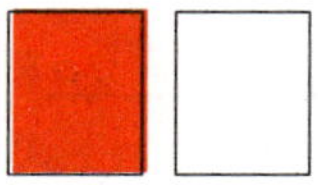	______	______	______
4.	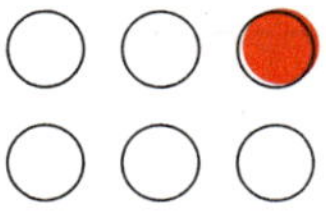	______	______	______
5.	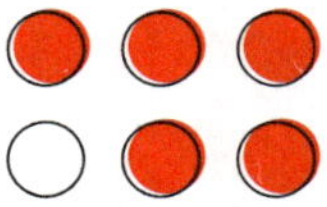	______	______	______

Lesson 2 Unit 3

REPRESENTING PARTS OF A WHOLE

Fractions can be used to show any number of equal parts in a shape. Fractions can also describe members of a group of any size.

$\frac{5}{8}$ 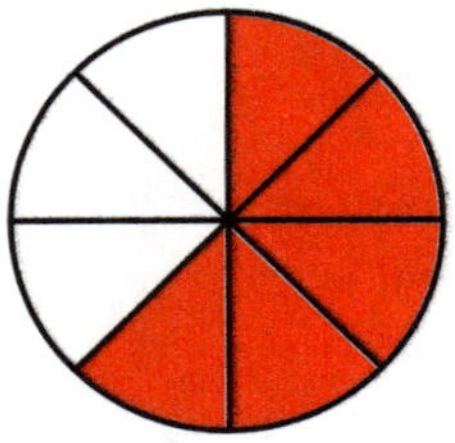

$\frac{2}{5}$ 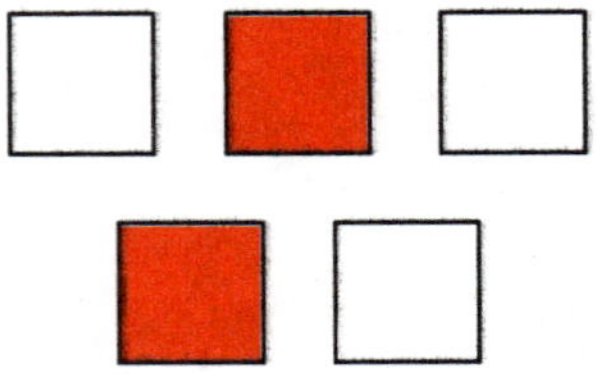

A Each of these drawings can be described with a fraction. The denominator is given. Count the shaded parts to write the numerator.

1. 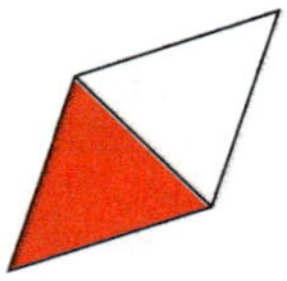 $\frac{\quad}{2}$

2.  $\frac{\quad}{16}$

B Write a fraction to show how many shaded parts are in each.

1.  ______________

2.  ______________

3. 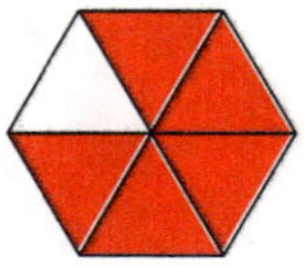 ______________

4. 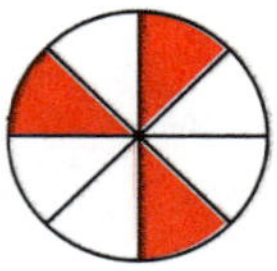 ______________

5. 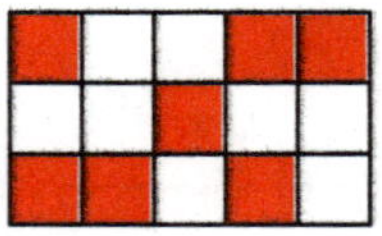 ______________

6. 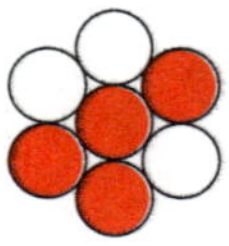 ______________

7. 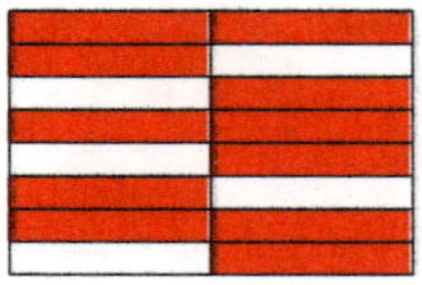 ______________

8. 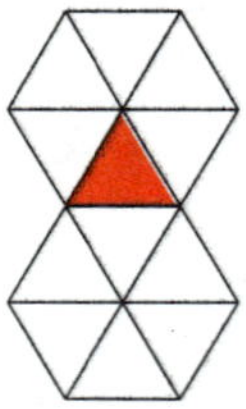 ______________

C Shade each drawing to show the fraction.

1. $\frac{1}{8}$

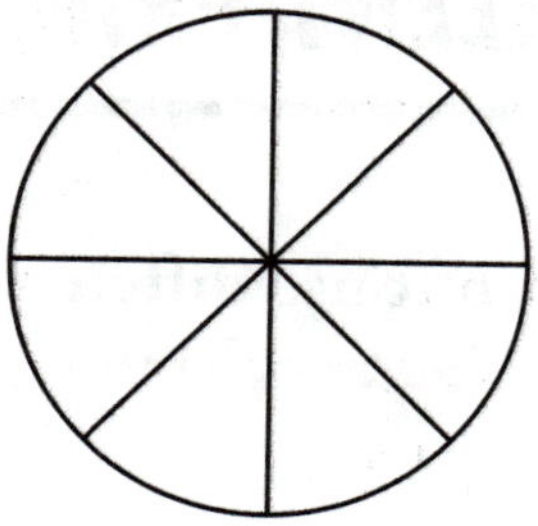

2. $\frac{2}{3}$

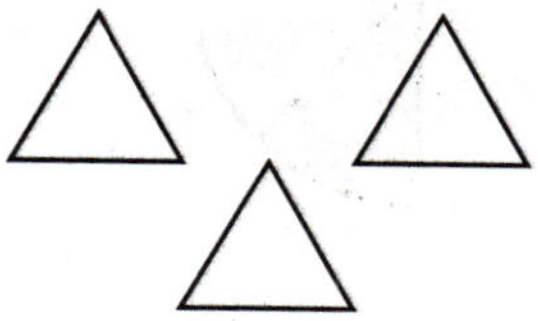

3. $\frac{3}{4}$

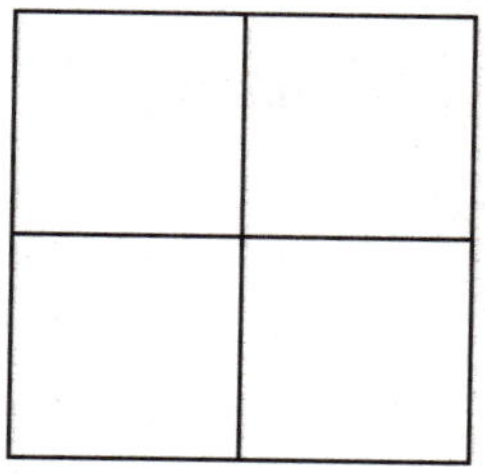

4. $\frac{5}{12}$

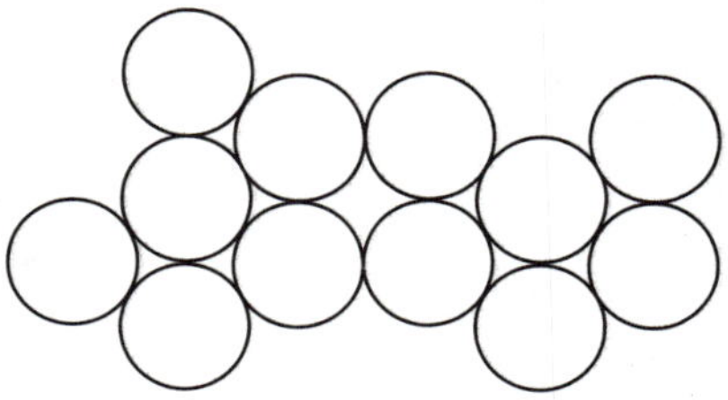

D Make your own drawings to represent each of these fractions. Divide each drawing into the number of equal parts given in the denominator. Shade in the number of equal parts given in the numerator.

1. $\frac{1}{5}$

2. $\frac{3}{4}$

3. $\frac{1}{6}$

4. $\frac{3}{10}$

5. $\frac{7}{8}$

6. $\frac{4}{9}$

UNDERSTANDING EQUIVALENT FRACTIONS

Fractions that show the same amount but use different numbers are equivalent fractions. ***Equivalent fractions*** are equal in value.

$\frac{1}{2}$ is the same as $\frac{3}{6}$

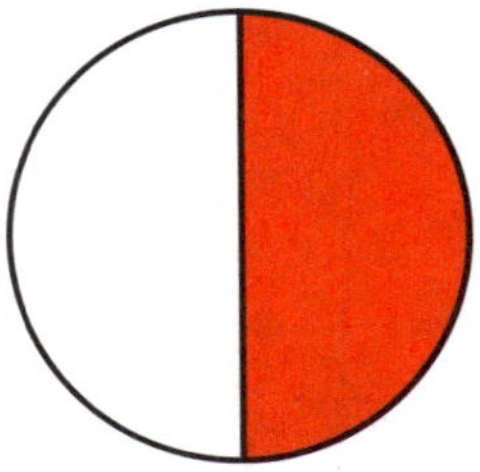

$\frac{1}{2}$

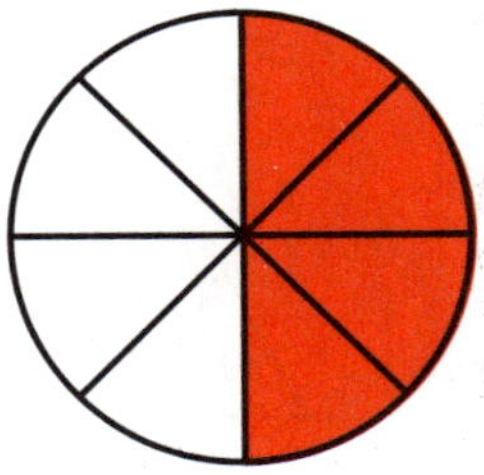

$\frac{3}{6}$

$\frac{1}{3}$ is the same as $\frac{3}{9}$

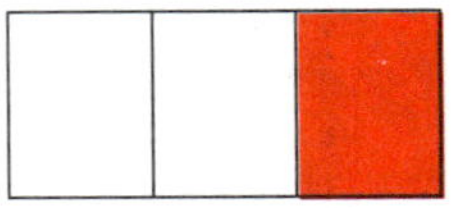

$\frac{1}{3}$

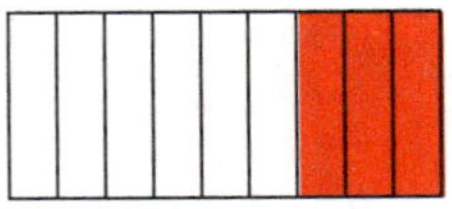

$\frac{3}{9}$

We use equivalent fractions every day when we tell time. There are 60 minutes in an hour.

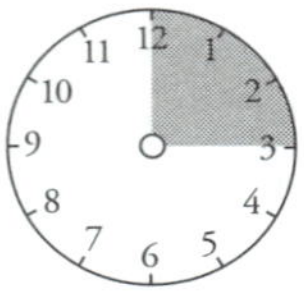

$\frac{15}{60} = \frac{1}{4}$ hour

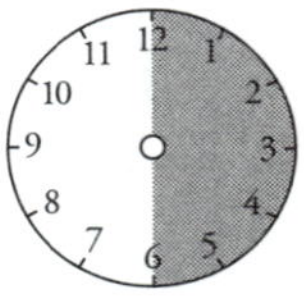

$\frac{30}{60} = \frac{1}{2}$ hour

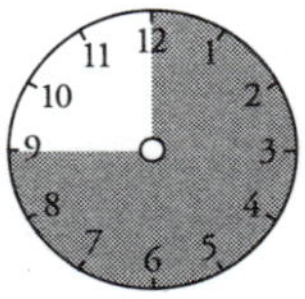

$\frac{45}{60} = \frac{3}{4}$ hour

A We use equivalent fractions in many kinds of common measurement. These pictures show a gas tank that holds 12 gallons. Fill in the missing numerators.

1. 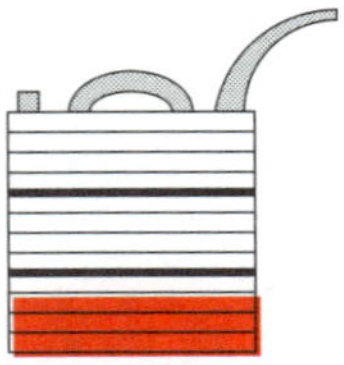
_____ tank = $\frac{3}{12}$

2. 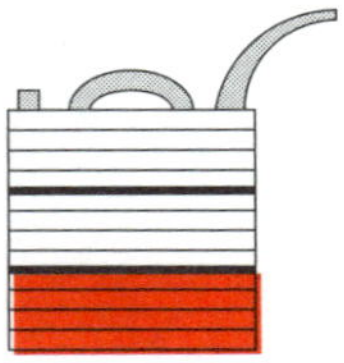
_____ tank = $\frac{}{12}$

3. 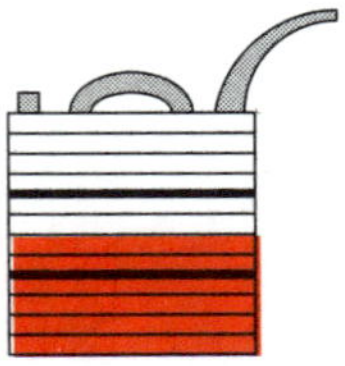
$\frac{1}{2}$ tank = $\frac{}{12}$

4. 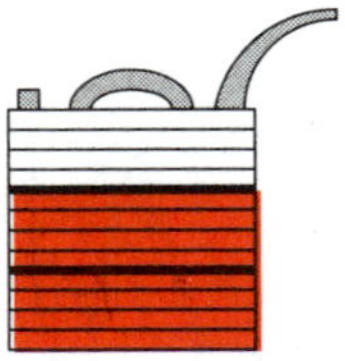
$\frac{2}{3}$ tank = $\frac{}{12}$

B Write equivalent fractions for each set of drawings.

1. 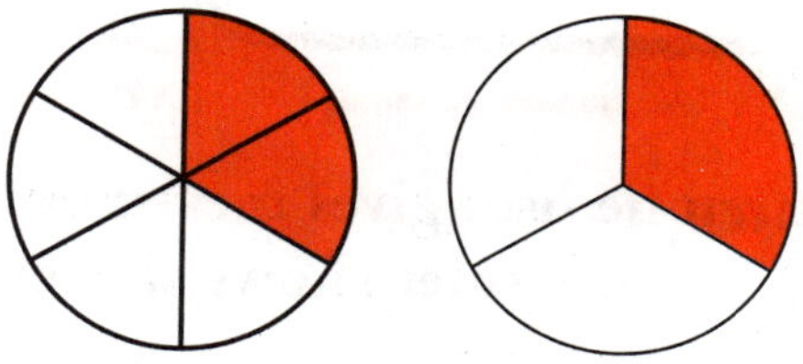

_______________ = _______________

2. 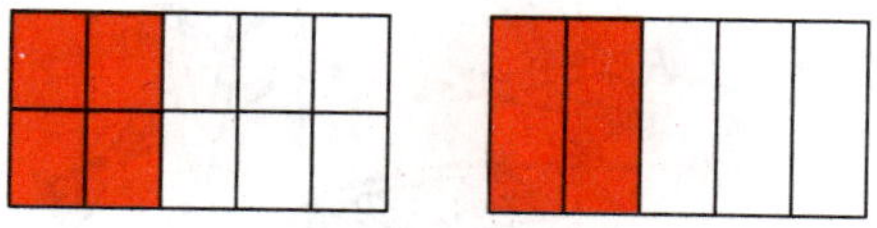

_______________ = _______________

3. 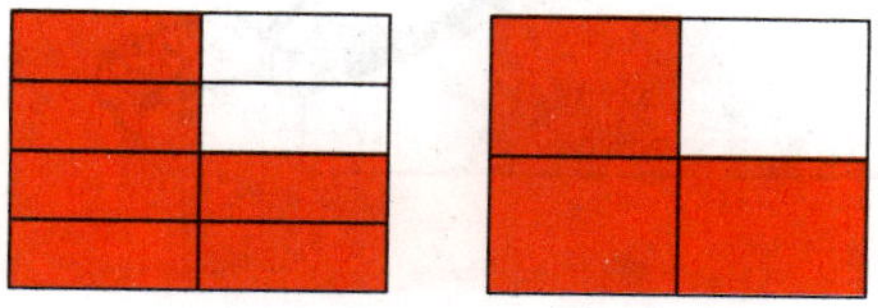

_______________ = _______________

C Shade the drawings to show equivalent fractions.

1. 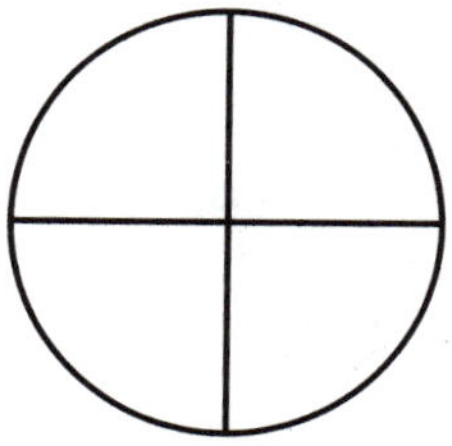 $\frac{2}{4}$ 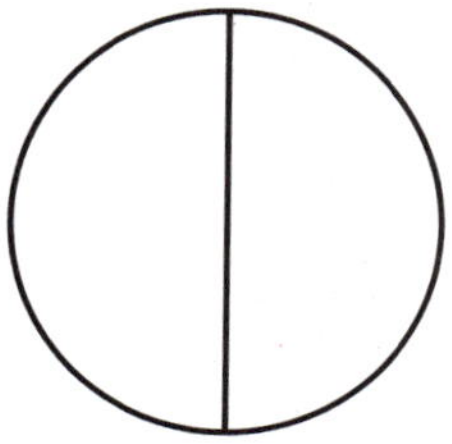 $\frac{1}{2}$

2. 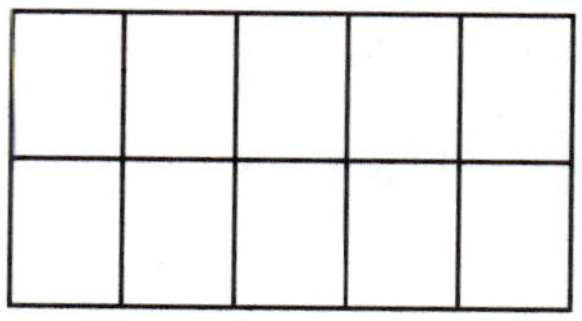 $\frac{8}{10}$ 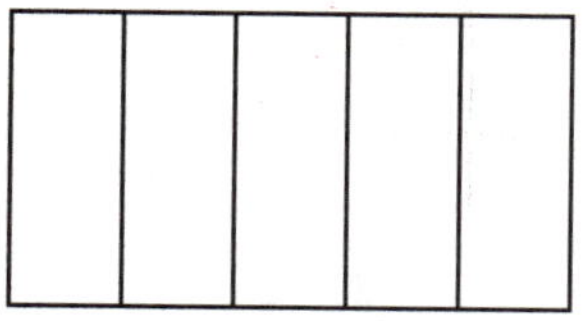 $\frac{4}{5}$

3. 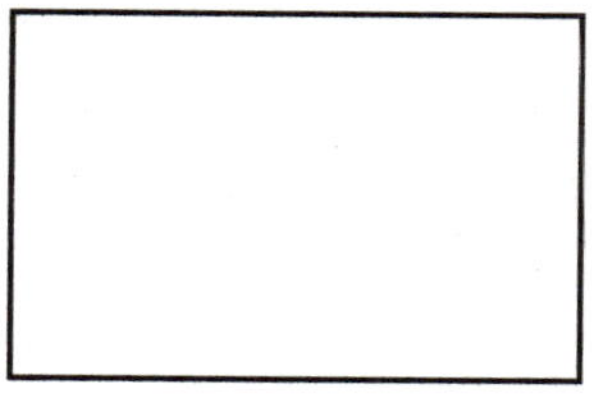 $\frac{6}{9}$ 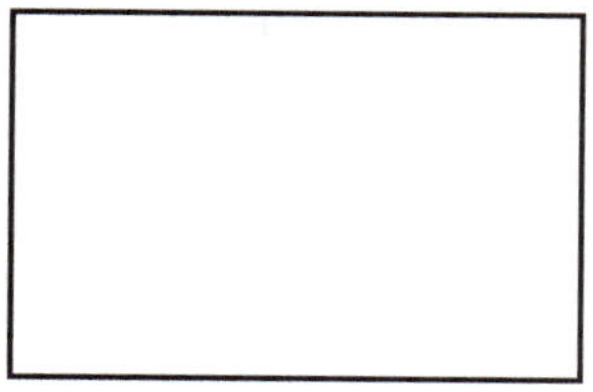 $\frac{2}{3}$

LESSON 4 UNIT 3

FINDING EQUIVALENT FRACTIONS

Jim is building a table. The pattern he uses gives measurements in half, quarter, and eighth inches. Jim's ruler shows sixteenths.

A Use these rulers to help Jim find equivalent fractions.

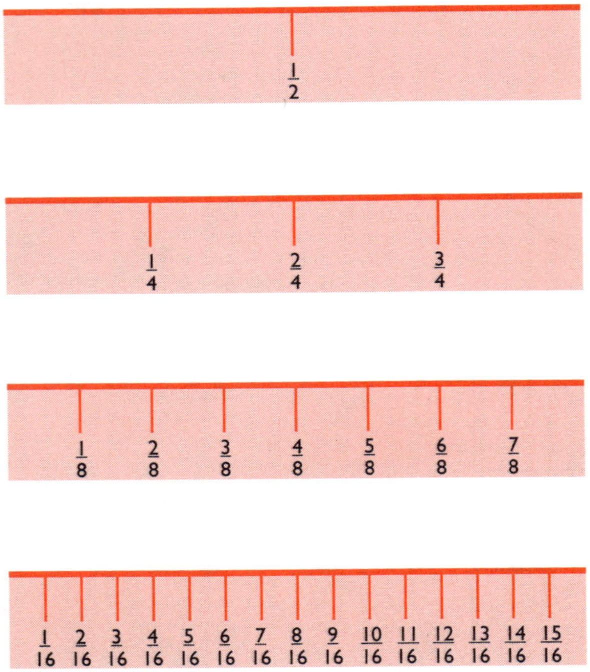

1. _____ $= \frac{2}{4} = \frac{4}{8} = \frac{8}{16}$

2. _____ $= \frac{2}{8} = \frac{\quad}{16}$

3. _____ $= \frac{6}{8} = \frac{\quad}{16}$

4. $\frac{3}{8} = \frac{\quad}{16}$

5. $\frac{5}{8} = \frac{\quad}{16}$

6. $\frac{\quad}{8} = \frac{14}{16}$

You can find equivalent fractions by multiplying. Multiply the numerator and denominator by the same number.

Example 1

$$\frac{2}{4} \times \frac{2}{2} = \frac{4}{8} \quad \text{so} \quad \frac{2}{4} = \frac{4}{8}$$

Example 2

$$\frac{3}{5} \times \frac{4}{4} = \frac{12}{20} \quad \text{so} \quad \frac{3}{5} = \frac{12}{20}$$

B Fill in the missing numbers for these equivalent fractions.

1. $\frac{1}{8} \times \frac{2}{2} = \frac{\underline{\quad}}{16}$ so $\frac{1}{8} = \frac{\underline{\quad}}{16}$

2. $\frac{3}{5} \times \frac{3}{3} = \frac{9}{\underline{\quad}}$ so $\frac{3}{5} = \frac{9}{\underline{\quad}}$

3. $\frac{1}{4} \times \frac{5}{5} = \frac{\underline{\quad}}{20}$ so _____ = _____

You can also find equivalent fractions by dividing. Divide the numerator and denominator by a common factor.

Example 1

$$\frac{6}{16} \div \frac{2}{2} = \frac{3}{8} \quad \text{so} \quad \frac{6}{16} = \frac{3}{8}$$

Example 2

$$\frac{12}{21} \div \frac{3}{3} = \frac{4}{7} \quad \text{so} \quad \frac{12}{21} = \frac{4}{7}$$

C Fill in the missing numbers and operations signs for these equivalent fractions.

1. $\frac{20}{25} \quad \frac{5}{5} = \frac{4}{\underline{\quad}}$ so $\frac{20}{25} = \frac{4}{\underline{\quad}}$

2. $\frac{10}{12} \quad \frac{2}{2} = \frac{\underline{\quad}}{6}$ so $\frac{10}{12} = \frac{\underline{\quad}}{6}$

3. $\frac{27}{30} \quad \frac{3}{3} = \frac{9}{\underline{\quad}}$ so $\frac{27}{30} = \underline{\quad}$

LESSON 5 UNIT 3

SIMPLIFYING FRACTIONS

To express a fraction in its simplest form, divide its numerator and denominator by their Greatest Common Factor (GCF).

Example: Express 18/24 in simplest form.

Factors of 18 are: 1, 2, 3, **6**, 9, 18
Factors of 24 are: 1, 2, 3, 4, **6**, 8, 12, 24

The Greatest Common Factor of 18 and 24 is 6.

$$\frac{18}{24} \div \frac{6}{6} = \frac{3}{4}$$

REMEMBER

Remember: A fraction is in its simplest form when the GCF of its numerator and denominator is 1.

A Express each fraction in simplest form. Show your work.

1. $\frac{6}{54}$
2. $\frac{75}{90}$
3. $\frac{110}{156}$
4. $\frac{24}{40}$
5. $\frac{180}{270}$
6. $\frac{90}{360}$
7. $\frac{63}{81}$
8. $\frac{56}{80}$
9. $\frac{8}{9}$
10. $\frac{40}{50}$
11. $\frac{9}{36}$
12. $\frac{2}{3}$

B Choose the best answer to each question.

1. Which fraction is in simplest form?

   A $\frac{4}{8}$ C $\frac{4}{5}$

   B $\frac{7}{21}$ D $\frac{2}{24}$

2. Which fraction is in simplest form?

   A $\frac{2}{5}$ C $\frac{2}{6}$

   B $\frac{3}{18}$ D $\frac{2}{20}$

3. Which fraction is NOT in simplest form?

   A $\frac{13}{17}$ C $\frac{4}{13}$

   B $\frac{24}{25}$ D $\frac{6}{30}$

4. Which fraction is NOT in simplest form?

   A $\frac{4}{5}$ C $\frac{4}{7}$

   B $\frac{5}{16}$ D $\frac{5}{15}$

5. In which pair of fractions are both in simplest form?

   A $\frac{5}{6}$, $\frac{7}{8}$ C $\frac{3}{4}$, $\frac{4}{8}$

   B $\frac{10}{12}$, $\frac{11}{13}$ D $\frac{12}{20}$, $\frac{13}{15}$

6. In which pair of fractions are both in simplest form?

   A $\frac{1}{3}$, $\frac{3}{9}$ C $\frac{3}{5}$, $\frac{3}{6}$

   B $\frac{1}{4}$, $\frac{5}{9}$ D $\frac{6}{30}$, $\frac{7}{30}$

In an improper fraction, the numerator is larger than the denominator. Improper fractions can be changed to mixed numbers by dividing.

Example: $\frac{19}{6}$ $\quad 6\overline{)19} = 3$, 18, remainder 1 $\quad 3\frac{1}{6}$

C Change these improper fractions to mixed numbers. Express the fractions in simplest form. Show your work.

1. $\frac{39}{6}$

2. $\frac{10}{8}$

3. $\frac{45}{10}$

4. $\frac{22}{5}$

5. $\frac{51}{6}$

6. $\frac{26}{10}$

Lesson 6 · Unit 3 — TEST TRACK

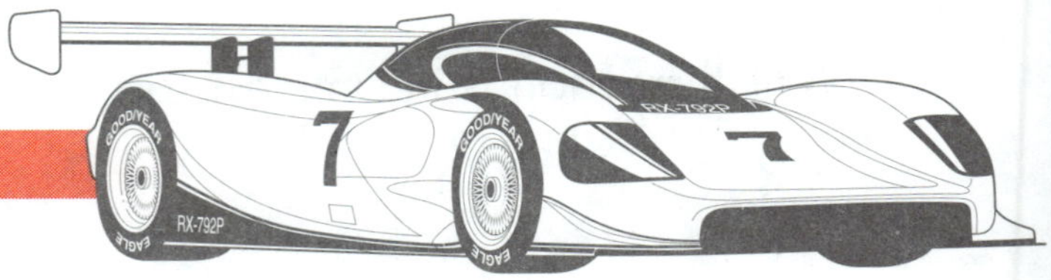

A Complete the number sentences to make equivalent fractions.

1. $\frac{4}{7} = \frac{}{21}$
2. $\frac{3}{9} = \frac{}{18}$
3. $\frac{2}{3} = \frac{}{15}$
4. $\frac{4}{5} = \frac{}{100}$
5. $\frac{4}{10} = \frac{}{5}$
6. $\frac{3}{9} = \frac{}{3}$
7. $\frac{5}{20} = \frac{}{4}$
8. $\frac{6}{36} = \frac{}{6}$
9. $\frac{2}{7} = \frac{6}{}$
10. $\frac{4}{11} = \frac{12}{}$
11. $\frac{18}{27} = \frac{2}{}$
12. $\frac{13}{26} = \frac{1}{}$

B Write = or ≠ to complete each number sentence.

1. $\frac{2}{7} \bigcirc \frac{6}{21}$
2. $\frac{4}{9} \bigcirc \frac{8}{18}$
3. $\frac{3}{5} \bigcirc \frac{6}{15}$
4. $\frac{12}{21} \bigcirc \frac{6}{7}$
5. $\frac{2}{11} \bigcirc \frac{6}{22}$
6. $\frac{6}{7} \bigcirc \frac{42}{49}$
7. $\frac{2}{3} \bigcirc \frac{5}{8}$
8. $\frac{3}{4} \bigcirc \frac{9}{10}$

REMEMBER

= means *equals*

≠ means *does not equal*

C Shade $\frac{1}{3}$ of each of the figures below.

1.

2.

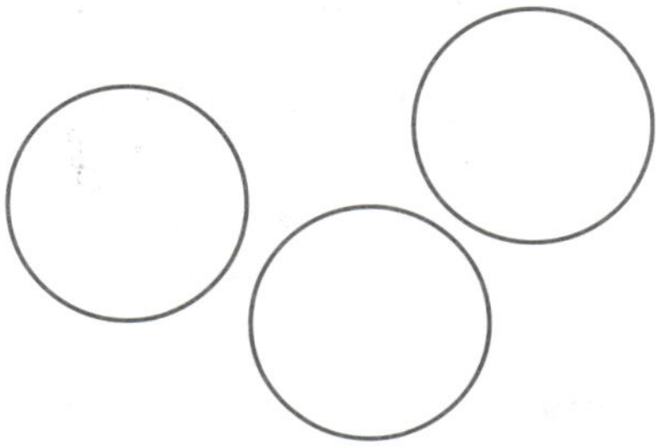

3.

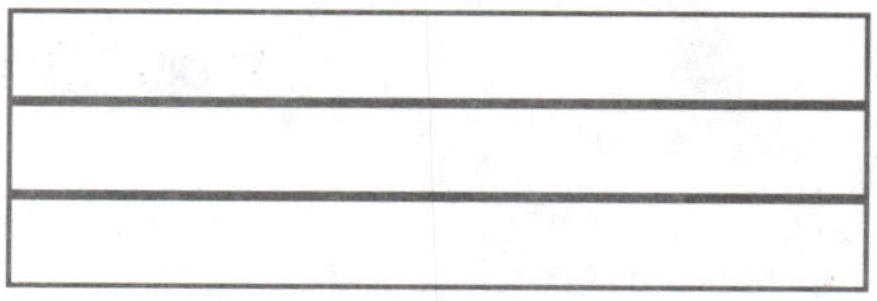

D Choose the best answer to each question.

1. Two ways to find equivalent fractions are—

Ⓐ addition and subtraction

Ⓑ addition and multiplication

Ⓒ multiplication and division

Ⓓ subtraction and division

2. Another way to express $\frac{2}{5}$ is —

Ⓐ five halves

Ⓑ five out of two

Ⓒ one fifth

Ⓓ two out of five

3. Joe bought a dozen cookies to share with friends. Four of them were eaten at lunch. Which fraction represents the remaining cookies?

Ⓐ $\frac{2}{3}$ Ⓑ $\frac{1}{3}$ Ⓒ $\frac{4}{12}$ Ⓓ $\frac{1}{12}$

4. Joe's dad ate two of the remaining cookies at dinner. Which fraction represents the cookies that had been eaten at lunch and dinner?

Ⓐ $\frac{1}{4}$ Ⓑ $\frac{1}{2}$ Ⓒ $\frac{5}{12}$ Ⓓ $\frac{2}{3}$

CHANGING FRACTIONS TO DECIMALS

Every fraction has an equivalent decimal. Fractions with powers of ten in the denominator are easy to change into decimals.

$\frac{6}{10} = 0.6$ $\frac{43}{100} = 0.43$ $\frac{755}{1000} = 0.755$

A Write the equivalent decimal for each fraction below.

1. $\frac{1}{10}$ = ______
2. $\frac{65}{100}$ = ______
3. $\frac{835}{1000}$ = ______
4. $\frac{5}{10}$ = ______
5. $\frac{75}{100}$ = ______
6. $\frac{1}{1000}$ = ______

You can change other fractions into decimals by dividing. Divide the numerator by the denominator.

Here are two ways to change $\frac{3}{4}$ into a decimal.

Pencil and paper method:

```
   0.75
4)3.00
  2 8
    20
    20
     0
```

Think of $\frac{3}{4}$ as a division problem. $\frac{3}{4}$ means three divided by four.

Divide. The quotient is the equivalent decimal.

Calculator method:

Press these keys in this order: [3] [÷] [4] [=]

B Use division to change each fraction into a decimal. Show your work.

1. $\frac{1}{4}$ $4\overline{)1.0}$ $\frac{1}{4}$ = __________

2. $\frac{1}{5}$ $5\overline{)1.0}$ $\frac{1}{5}$ = __________

3. $\frac{2}{5}$ $5\overline{)2.0}$ $\frac{2}{5}$ = __________

4. $\frac{4}{5}$ $5\overline{)4.0}$ $\frac{4}{5}$ = __________

5. $\frac{12}{20}$ $20\overline{)12.0}$ $\frac{12}{20}$ = __________

C Use a calculator to change these fractions into decimals.

1. $\frac{3}{15}$ = __________

2. $\frac{7}{28}$ = __________

3. $\frac{3}{12}$ = __________

4. $\frac{15}{60}$ = __________

5. $\frac{36}{40}$ = __________

6. $\frac{18}{24}$ = __________

LESSON 2 UNIT 4

PERCENTS

Decimals can be easily changed into percents. For decimals like the ones below, take out the decimal point, then add the percent sign.

0.03 = 3%   0.35 = 35%   0.75 = 75%   1.25 = 125%

A Change each decimal to a percent.

1. 0.43 = ________%   2. 0.22 = ________%   3. 1.52 = ________%

If the decimal has only a digit in the tenths place, move the decimal point two places to the right and add the %.

0.5 = 50%   0.8 = 80%

B Change each decimal to a percent.

1. 0.6 = ________%   2. 0.2 = ________%   3. 2.3 = ________%

Some decimals have more than two digits to the right of the decimal point. For these, place a new decimal point after the hundredths place when you write the percent.

0.375 = 37.5%   0.5056 = 50.56%

C Change each decimal to a percent.

1. 0.471 = ________%   2. 0.8991 = ________%   3. 1.442 = ________%

Percent means "hundredths." Every percent is equal to a fraction with 100 in the denominator.

5/100 = 5% 73/100 = 73%

D Change each fraction to a percent.

1. $\frac{20}{100}$ = ________%
2. $\frac{62}{100}$ = ________%
3. $\frac{35}{100}$ = ________%
4. $\frac{50}{100}$ = ________%

If a fraction has 10 in the denominator, you can add a zero to the numerator and denominator.

$$\frac{3}{10} = \frac{3}{10} \times \frac{10}{10} = \frac{30}{100} = 30\%$$

E Change each fraction to percent.

1. $\frac{4}{10} = \frac{40}{100}$ = ________%
2. $\frac{9}{10} = \frac{\quad}{100}$ = ________%
3. $\frac{7}{10} = \frac{\quad}{100}$ = ________%
4. $\frac{12}{10} = \frac{120}{100}$ = ________%

F Complete the number sentences below.

1. ________ = 0.5 = ________%
2. $\frac{4}{10}$ = ________ = ________%
3. ________ = 0.25 = ________%
4. $\frac{6}{10}$ = ________ = ________%

LESSON 3 UNIT 4

RATIONAL NUMBERS

The chart below shows three ways of naming the same amount. The expression we usually use is shown in bold type for each example.

fraction	decimal	percent
$\frac{3}{4}$ of a dollar	**0.75 of a dollar (usually written $.75)**	75% of a dollar
$\frac{1}{2}$ of a loaf of bread	0.5 of a loaf	50% of a loaf
$\frac{9}{10}$ questions correct	0.9 correct	**90% correct**

For every fraction, there is an equivalent decimal and an equivalent percent.

fraction	decimal	percent
$\frac{1}{10}$	0.1	10%
$\frac{2}{10}$	0.2	20%
$\frac{3}{10}$	0.3	30%
$\frac{4}{10}$	0.4	40%
$\frac{5}{10}$	0.5	50%
$\frac{6}{10}$	0.6	60%
$\frac{7}{10}$	0.7	70%
$\frac{8}{10}$	0.8	80%
$\frac{9}{10}$	0.9	90%

fraction	decimal	percent
$\frac{1}{4}$	0.25	25%
$\frac{1}{2}$	0.5	50%
$\frac{3}{4}$	0.75	75%

fraction	decimal	percent
$\frac{1}{5}$	0.2	20%
$\frac{2}{5}$	0.4	40%
$\frac{3}{5}$	0.6	60%
$\frac{4}{5}$	0.8	80%

A Write a fraction, decimal, and percent for the shaded parts of each drawing.

		fraction	decimal	percent
1.	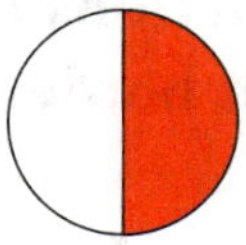	______	__________	__________
2.	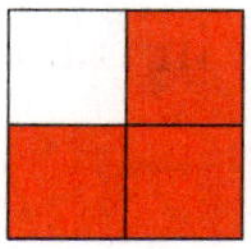	______	__________	__________
3.	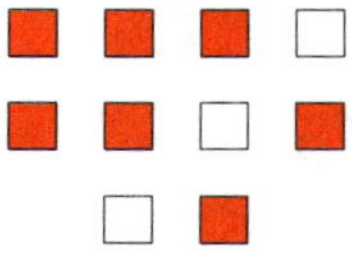	______	__________	__________
4.	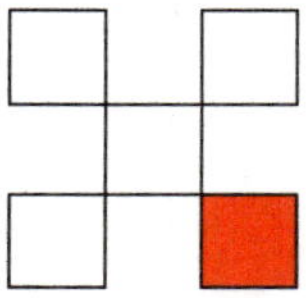	______	__________	__________
5.	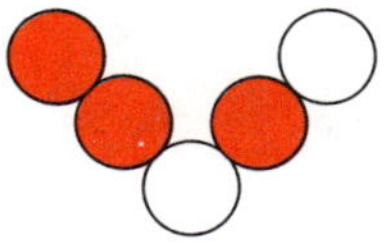	______	__________	__________
6.	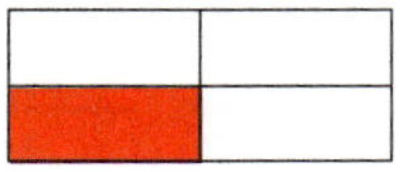	______	__________	__________

B Make a drawing to show each fraction, decimal, or percent.

1. $\frac{2}{5}$
2. $\frac{3}{4}$
3. 0.5
4. 40%
5. 60%
6. 80%

LESSON 4 UNIT 4

SOLVING WORD PROBLEMS

Word problems sometimes ask you to change fractions to decimals or percents. These problems look much harder than they are.

Janna works at the supermarket. She works at various jobs every day. On Tuesday, Janna worked $\frac{3}{4}$ hour in the meat department. She records her hours in fractions, but her supervisor needs to report them in decimals.

To change $\frac{3}{4}$ to a decimal, divide.

$$
\begin{array}{r}
0.75 \\
4\overline{)3.00} \\
\underline{2\;8} \\
2\;0 \\
\underline{2\;0} \\
0
\end{array}
\qquad \frac{3}{4} = 0.75
$$

REMEMBER
The denominator becomes the divisor and the numerator becomes the dividend. Add a zero and decimal point to the dividend.

A Solve each problem. Follow the steps.

1. The music room has 100 chairs. There are 35 people in the room, and each of them is sitting in a chair. What percent of the chairs are filled?

   Step One: Write a fraction. ____________

   Step Two: Convert to a decimal. Because the denominator is 100 you don't have to divide. Write the numerator with a percent sign after it.

   Answer: ____________

2. An advertisement says that seven out of ten people prefer Bright White toothpaste. What percent prefer Bright White?

Step One: Write a fraction. ____________

Step Two: Convert to a percent. Because the denominator is 10, you don't have to divide. Add a zero to the numerator and write a percent sign after it.

Answer: ____________%

---

3. In 20 grams of mystery meat there are 6 grams of fat. What percent of the meat is fat?

Step One: Write a fraction. ____________

Step Two: Convert to a decimal by dividing.

$$20\overline{)6.0} = 0.3$$

Step Three: Change the decimal to a percent. Add a zero if there is only one digit in the decimal.

Answer: ____________

---

4. A pie is divided into 5 equal pieces. Three pieces have been eaten. What percent are left?

Step One: Draw the pie. Shade in the three missing pieces.

Step Two: Write a fraction to show how many are left. ____________

Step Three: Convert to a decimal. Show your work.

Step Four: Change the decimal to a percent.

Answer: ____________

LESSON 5 UNIT 4

TEST TRACK

A Answer the following questions. Show your work.

1. Jim has 10 math problems to complete for homework. He's already finished 6 of them. What percent of his math homework remains to be completed?

2. In Ari's drama class, 5/8 of the students are girls. Write this fraction as a decimal.

3. Three quarters of the people attending the dinner have requested chicken. What percent of the people have requested something other than chicken?

4. The Second National Bank charges 9.8% interest on used car loans. Write this percent as a decimal.

5. Frank jogged 3.375 miles on Thursday. Express this decimal as a mixed number.

B Choose the best answer to each of the following questions.

1. Lon has saved 30% of the money she needs for her trip. Which fraction represents how much she has saved?

Ⓐ $\frac{1}{3}$ Ⓑ $\frac{3}{5}$ Ⓒ $\frac{3}{10}$ Ⓓ $\frac{3}{100}$

2. With 2 km of the 5 km race still ahead of him, Stacey had to stop to rest. What percent of the race had he completed when he stopped?

Ⓐ 35% Ⓑ 40% Ⓒ 50% Ⓓ 60%

3. Tom drank 4 cans of soda from the 12 pack. Which fraction represents the amount he drank?

Ⓐ $\frac{2}{3}$ Ⓑ $\frac{1}{3}$ Ⓒ $\frac{1}{2}$ Ⓓ $\frac{1}{4}$

4. Angie got 18 out of 20 math test questions correct. Her teacher wrote 0.9 at the top of her paper. What percent score did Angie get?

Ⓐ 9% Ⓑ 90% Ⓒ 18% Ⓓ 99%

5. Tina's gas tank was 7/8 full when she returned the car to her mom. What decimal represents the amount of gas in the tank?

Ⓐ 0.750 Ⓑ 0.785 Ⓒ 0.625 Ⓓ 0.875

C Express the shaded region of the grid in three ways.

1.

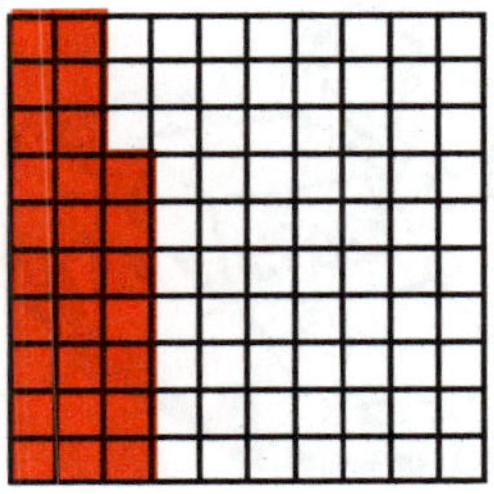

fraction decimal percent

LESSON 1 UNIT 5

COMPARING WHOLE NUMBERS

Andreas is shopping for a used car. He has kept this list of prices for cars that he likes, and wants to compare them to find the best bargain.

To compare prices, he lists the prices in columns showing place values. Then he reads the columns from the left.

	thousands	hundreds	tens	ones
Wagon	1	2	9	5
Sedan	1	2	7	5
SUV	2	0	2	5
Van	1	7	9	9
Pickup	2	0	9	5

Which is the highest price? Look at the thousands place. The prices with a 2 are higher than those with a 1. For these prices only, look at the hundreds place. Both prices have a 0 in the hundreds place. Look at the tens place. 9 is higher than 2, so the highest price is the pickup at $2,095.

Which is the lowest price? Look at the thousands place. The prices with a 1 are lower than those with a 2. For these prices only, look at the hundreds place. 2 is lower than 7, so put another X next to the prices that have two in the hundreds place. For these prices only, look at the tens place. 7 is lower than 9, so the sedan is lowest at $1,275.

A List the numbers in the place value chart from greatest to least.

1. 156 2. 267 3. 256

4. 1,345 5. 1,090 6. 2,004

thousands	hundreds	tens	ones

Inequality signs are often used to compare numbers.

10 > 9 — 10 <u>is greater than</u> 9

9 < 10 — 9 <u>is less than</u> 10

The pointed end of the sign always points to the lesser number. The open end of the sign always faces the greater number.

"greater than" means "larger than"

"less than" means "smaller than"

B Use inequality signs to compare each pair of numbers. Make a place value chart if it helps you find the greater number.

1. 16 (>) 15
2. 5 ( ) 8
3. 64 ( ) 34
4. 12 ( ) 241
5. 1,799 ( ) 1,977
6. 2,025 ( ) 1,475

WORK SPACE

LESSON 2 UNIT 5

COMPARING DECIMALS

Decimals can also be compared using a place value chart. Compare these decimals:

0.357 0.027 0.92 0.0915 0.308 0.9

First, put the decimals into a place value chart. Then add zeros for any empty places.

You can add any number of zeros to the end of a decimal, without changing its meaning.

0.6 = 0.60 = 0.6000000

tenths	hundredths	thousandths	ten thousandths
.3	5	7	0
.0	2	7	0
.9	2	0	0
.0	9	1	5
.3	0	8	0
.9	0	0	0

Which decimal is the greatest? Starting at the left, look at the tenths place. The greatest digit is 9. For these decimals only, look at the hundredths place. The greatest digit is 2, so 0.92 is the greatest decimal.

Which decimal is the least? Look at the tenths place. The least digit is 0. For these decimals only, look at the hundredths place. The least digit is 2, so 0.027 is the least decimal.

A

Circle the greater decimal in each pair. Remember to look at place values, and add zeros to empty places.

1. 0.81 0.71
2. 0.654 0.66
3. 0.013 0.01
4. 0.995 0.959
5. 0.034 0.304
6. 0.057 0.077

B Use inequality signs to compare each pair of decimals. Make a place value chart if it helps you find the larger number.

1. 0.4 (<) 0.6
2. 0.25 ◯ 0.2
3. 0.1 ◯ 0.01
4. 0.51 ◯ 0.53
5. 0.361 ◯ 0.301
6. 0.78 ◯ 0.79

C Circle the greatest decimal in each set. Choose at least two of the numbers and write a number sentence comparing them. Remember to look at place values, and add zeros to empty places.

1. 0.76 0.75 0.69 ______________________
2. 0.019 0.02 0.01 ______________________
3. 0.9 091 0.01 ______________________
4. 0.98 0.99 0.97 ______________________

D Write a number to answer each question.

1. What number is 3 tenths greater than 54.315? __________
2. What number is 6 hundredths less than 4.283? __________
3. What number is 4 hundredths more than 1,437.133? __________
4. What number is 8 tenths less than 456.873? __________
5. What number is 1 greater than 0.034? __________

Lesson 3 · Unit 5

COMPARING FRACTIONS

Like fractions have the same denominator. Unlike fractions have different denominators.

$\frac{2}{5}$ and $\frac{3}{5}$ are like fractions.

$\frac{2}{5}$ and $\frac{3}{7}$ are unlike fractions.

To compare like fractions, look at their numerators.

$\frac{3}{5} > \frac{2}{5}$

To compare unlike fractions, express each using their least common denominator. Then, look at their numerators.

Example: Which is greater? $\frac{2}{5}$ or $\frac{3}{7}$?

Step 1: Find the Least Common Denominator

Multiples of 5: 5, 10, 25, 20, 25, 30, 35 . . .
Multiples of 7: 7, 14, 21, 28, 35 . . .
35 is the least common multiple of 5 and 7.

HINT
The least common denominator is the least common multiple of the denominators.

Step 2: Find an equivalent fraction for each with the denominator 35.

$\frac{2 \times 7}{5 \times 7} = \frac{14}{35}$

$\frac{3 \times 5}{7 \times 5} = \frac{15}{35}$

Step 3: Compare numerators.

$\frac{15}{35} > \frac{14}{35}$ so $\frac{3}{7} > \frac{2}{5}$

A List the multiples for each pair of denominators. Circle the least common multiple.

1. $\frac{12}{14}, \frac{7}{8}$

____

____

3. $\frac{2}{3}, \frac{4}{7}$

____

____

2. $\frac{17}{30}, \frac{3}{5}$

____

____

4. $\frac{14}{15}, \frac{11}{12}$

____

____

B Express both fractions using their Least Common Denominator.

1. $\frac{12}{14}, \frac{7}{8}$

3. $\frac{2}{3}, \frac{4}{7}$

2. $\frac{17}{30}, \frac{3}{5}$

4. $\frac{14}{15}, \frac{11}{12}$

C Compare the fractions. Show your work.

WORK SPACE

1. $\frac{4}{7}$ ◯ $\frac{1}{2}$

4. $\frac{2}{3}$ ◯ $\frac{8}{11}$

2. $\frac{3}{10}$ ◯ $\frac{1}{3}$

5. $\frac{3}{4}$ ◯ $\frac{5}{6}$

3. $\frac{18}{24}$ ◯ $\frac{10}{18}$

6. $\frac{4}{9}$ ◯ $\frac{7}{12}$

LESSON 4 UNIT 5

COMPARING FRACTIONS

Another way to compare fractions is to use multiplication.
Which is greater, $\frac{2}{3}$ or $\frac{4}{5}$?

Multiply each denominator by the numerator of the other fraction. This makes a criss-cross pattern, or X. Compare the products.

5 x 2 = 10 3 x 4 = 12

$\frac{2}{3}$ $\frac{4}{5}$

12 is greater than 10.

$\frac{4}{5}$ is greater than $\frac{2}{3}$

$\frac{4}{5} > \frac{2}{3}$

A Criss-cross multiply and fill in the missing numbers.

1. 4 X 2 = 8 3 X 3 = 9

$\frac{2}{3}$ $\frac{3}{4}$

larger product ___9___

larger fraction ________

2. 5 X 2 = ________ 7 X 2 = ________

$\frac{2}{7}$ $\frac{2}{5}$

larger product ________

larger fraction ________

3. 4 X 1 = ________ 3 X 1 = ________

$\frac{1}{3}$ $\frac{1}{4}$

larger product ________

larger fraction ________

B Criss-cross multiply, then use the inequality sign to compare the fractions.

1. $2 \times 5 =$ ______ $8 \times 1 =$ ______ $\frac{5}{8}$ ◯ $\frac{1}{2}$

$\frac{5}{8} \times \frac{1}{2}$

2. $4 \times 2 =$ ______ $9 \times 1 =$ ______ $\frac{2}{9}$ ◯ $\frac{1}{4}$

$\frac{2}{9} \times \frac{1}{4}$

3. $5 \times 1 =$ ______ $2 \times 2 =$ ______ $\frac{1}{2}$ ◯ $\frac{2}{5}$

$\frac{1}{2} \times \frac{2}{5}$

C Use inequality signs to compare the fractions. Show your work.

1. $\frac{2}{7}$ ◯ $\frac{2}{5}$

2. $\frac{1}{3}$ ◯ $\frac{2}{9}$

3. $\frac{7}{8}$ ◯ $\frac{5}{6}$

4. $\frac{4}{5}$ ◯ $\frac{6}{7}$

5. $\frac{3}{4}$ ◯ $\frac{6}{7}$

6. $\frac{4}{7}$ ◯ $\frac{1}{8}$

LESSON 5 UNIT 5

USING A NUMBER LINE

A number line can be used to put numbers in order. Number lines can be divided into any size equal units.

Each interval on this number line stands for ten.

0 10 20 30 40 50 60 70 80 90 100

Fractions can also be shown on a number line.

Each interval on this number line stands for $\frac{1}{2}$.

$0 \quad \frac{1}{2} \quad 1 \quad 1\frac{1}{2} \quad 2 \quad 2\frac{1}{2} \quad 3 \quad 3\frac{1}{2} \quad 4 \quad 4\frac{1}{2} \quad 5$

Numbers like $1\frac{1}{2}$, $2\frac{1}{2}$, $3\frac{1}{2}$, $4\frac{1}{2}$, etc. are called mixed numbers.

Each interval on this number line stands for $\frac{1}{3}$.

$0 \quad \frac{1}{3} \quad \frac{2}{3} \quad 1 \quad 1\frac{1}{3} \quad 1\frac{2}{3} \quad 2 \quad 2\frac{1}{3} \quad 2\frac{2}{3} \quad 3$

A Fill in the missing numbers. Complete each statement.

1. 0 2 [ ] 6 [ ] 10 12 [ ] 16 18

Each interval on this number line stands for _____.

2. 0 3 6 [ ] 12 15 18 [ ] 24 27

Each interval on this number line stands for _____.

B Complete each statement about the number line below.

0 5 10 15 20 25 30 35 40 45 50 55 60

1. 7 is between 5 and 10.

2. 21 is between 20 and _____.

3. 43 is between _____ and 45 .

4. 18 is between _____ and _____.

5. 36 is between _____ and _____.

6. 59 is between _____ and _____.

7. Each interval on this number line stands for _____.

C Fill in the missing numbers.

1.

$0 \quad \frac{1}{5} \quad \frac{2}{5} \quad \square \quad \frac{4}{5} \quad 1 \quad 1\frac{1}{5} \quad \square \quad 1\frac{3}{5} \quad 1\frac{4}{5} \quad \square \quad 2\frac{1}{5} \quad 2\frac{2}{5}$

Each interval on this number line stands for _____.

2.

$0 \quad \frac{1}{7} \quad \frac{2}{7} \quad \frac{3}{7} \quad \square \quad \frac{5}{7} \quad \frac{6}{7} \quad \square \quad 1\frac{1}{7} \quad 1\frac{2}{7} \quad \square \quad \square \quad 1\frac{5}{7}$

Each interval on this number line stands for _____.

LESSON 6 UNIT 5

TEST TRACK

A Answer the following questions. Show your work.

McKinley School - Week 17	
Day	Number of Students in Attendance
Monday	643
Tuesday	622
Wednesday	608
Thursday	597
Friday	615

1. The table shows the number of students who attended school each day last week. Which list shows student attendance in order from LEAST to GREATEST?

   Ⓐ Thursday, Wednesday, Friday, Tuesday, Monday
   Ⓑ Thursday, Friday, Tuesday, Monday, Wednesday
   Ⓒ Monday, Tuesday, Friday, Wednesday, Thursday
   Ⓓ Wednesday, Friday, Tuesday, Monday, Thursday

2. Alex bought $2\frac{1}{2}$ pounds of steak, a $1\frac{2}{3}$ pound lobster, and $2\frac{1}{3}$ pounds of chicken for the cook-out. Which list shows the weights in order by type of meat?

   Ⓐ steak, chicken, lobster
   Ⓑ chicken, lobster, steak
   Ⓒ chicken, steak, lobster
   Ⓓ lobster, steak, chicken

3. In a recent poll, $\frac{2}{5}$ of those questioned were in favor of a resolution to fund park renovation. A little more than 20% of those polled had no opinion. Which could be the percent of those who were opposed?

   Ⓐ 10% Ⓑ 36% Ⓒ 40% Ⓓ 60%

4. Josh's doctor told him that, at 208 pounds, he needed to lose at least 25 pounds. The next time he went to the doctor, he had more than achieved his goal. Which could be Josh's new weight?

   Ⓐ 187 Ⓑ 178 Ⓒ 183 Ⓓ 225

5. The distance between Kent and Ridgeville is 16 miles. Which point best represents this distance?

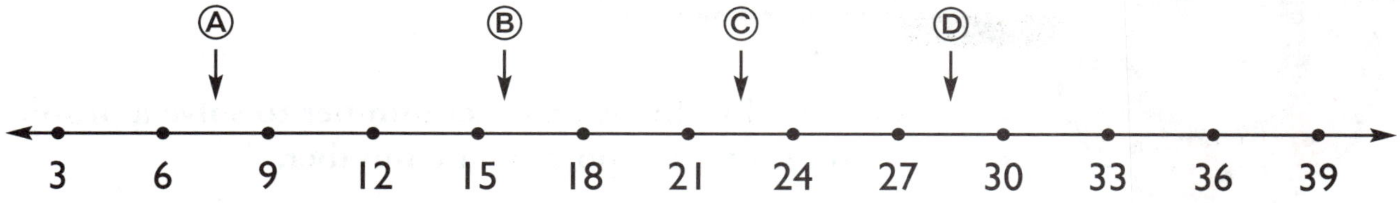

6. Which letter on the number line best represents $6\frac{1}{2}$?

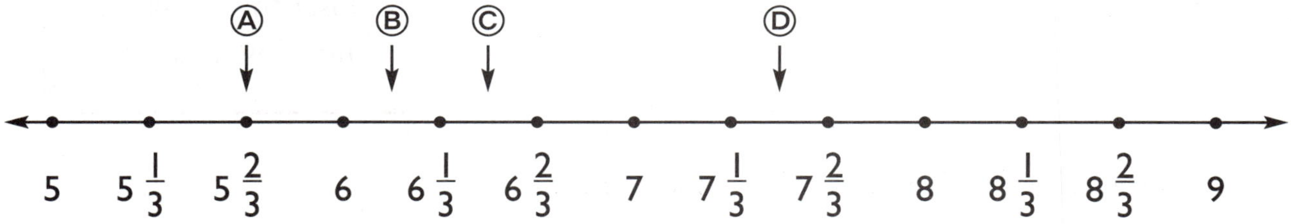

B Use inequality signs to compare the fractions. Show your work.

1. $\frac{7}{10}$ ◯ $\frac{3}{5}$

4. $\frac{1}{4}$ ◯ $\frac{6}{25}$

2. $\frac{6}{11}$ ◯ $\frac{12}{20}$

5. $\frac{3}{4}$ ◯ $\frac{23}{30}$

3. $\frac{1}{4}$ ◯ $\frac{4}{15}$

6. $\frac{4}{7}$ ◯ $\frac{5}{25}$

UNIT 6

LESSON 1

ROUNDING

If you don't need to know an exact number to solve a problem, you can round it to an approximate number.

About how many days are there in a month? Months can have 30, 31, 29, or 28 days.

Look at this number line.

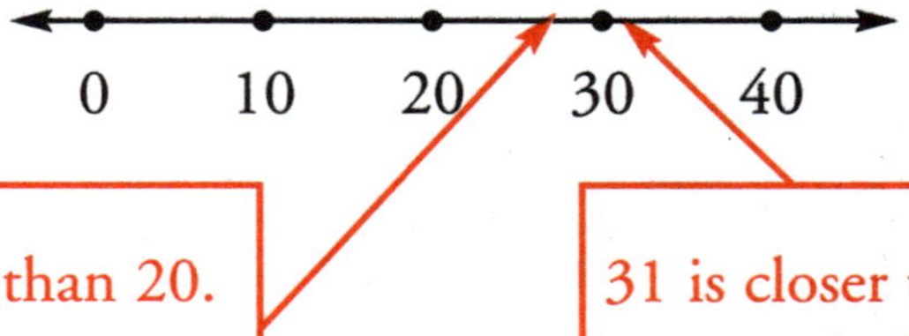

28 is closer to 30 than 20.
28 rounds up to 30.

31 is closer to 30 than 40.
31 rounds down to 30.

Answer: There are about 30 days in a month.

Rule for Rounding: 5, 6, 7, 8, 9 round up 4, 3, 2, 1 round down

You can round a number to any place value.

65,47**3** rounded to nearest ten:	65,470
65,4**7**3 rounded to nearest hundred:	65,500
65,**4**73 rounded to nearest thousand:	65,000
6**5**,473 rounded to nearest ten-thousand:	70,000

A Practice rounding.

Round to the nearest ten.

1. 564 ________
2. 1,272 ________
3. 89 ________
4. 67 ________

Round to the nearest hundred.

5. 372 ________
6. 5,647 ________
7. 290 ________
8. 2,612 ________

Round to the nearest thousand.

9. 7,827 ________ 11. 14,240 ________

10. 82,610 ________ 12. 6,419 ________

You can round decimals in almost the same way.

Round 0.261 to the nearest hundredth.

tenths	hundredths	thousandths
.2	6	1

1 is less than 5, so **round down** to 0.260
Decimals usually drop any 0's at the end,
so write 0.260 as 0.26

Round 0.261 to the nearest tenth.

tenths	hundredths	thousandths
.2	6	1

6 is greater than 5, so **round up** to 0.300 = 0.3

B Practice rounding.

Round to the nearest thousandth.

1. 0.7842 0.7840 = 0.784 3. 0.9496 ________ = ________

2. 0.1437 ________ = ________ 4. 0.4151 ________ = ________

Round to the nearest hundredth.

5. 0.654 ________ = ________ 7. 0.8764 ________ = ________

6. 0.6872 ________ = ________ 8. 0.9213 ________ = ________

Round to the nearest tenth.

9. 0.76 ________ = ________ 11. 0.91 ________ = ________

10. 0.065 ________ = ________ 12. 0.421 ________ = ________

LESSON 2 UNIT 6

ROUNDING MIXED NUMBERS

A mixed number contains a whole number and a fraction or decimal.

Examples of mixed numbers are: $2\frac{1}{2}$ 6.7 $45\frac{2}{3}$ $9\frac{1}{7}$ 18.87

Mixed numbers with **decimals** can be rounded in the same way as simple decimals. Just include the whole number place values when you study the number.

Round 18.87 to the nearest whole number.

tens	ones	tenths	hundredths
1	8	.8	7

8 is greater than 5, so **round up** to 19.00 = 19

A Complete the chart.

	rounded to the neatest ten	rounded to the nearest whole number	rounded to the nearest tenth
1. 7.18			
2. 26.56			
3. 52.74			
4. 65.48			
5. 81.9			
6. 4.581			
7. 30.02			
8. 11.65			

Mixed numbers with fractions can also be rounded to the nearest whole number. If the fraction is $\frac{1}{2}$ or greater, round up. If the fraction is less than $\frac{1}{2}$, round down.

Example: Which is greater, $2\frac{1}{2}$ or $2\frac{1}{3}$?

HINT
Criss-cross multiply to compare fractions.

$2 \times 1 = 2$ ← $\frac{1}{3}$ ✕ $\frac{1}{2}$ → $3 \times 1 = 3$

$\frac{1}{3} < \frac{1}{2}$ so, $2\frac{1}{3}$ rounds down to 2 and $2\frac{1}{2}$ rounds up to 3.

B **Compare each fraction. to $\frac{1}{2}$. Round the mixed number to the nearest whole number.**

1. $6\frac{5}{6}$ — $2 \times 5 = 10$ ← $\frac{5}{6}$ ✕ $\frac{1}{2}$ → $6 \times 1 = 1$ — $\frac{5}{6}$ (>) $\frac{1}{2}$ so round to 7

2. $10\frac{4}{5}$ — $2 \times 4 =$ ___ ← $\frac{4}{5}$ ✕ $\frac{1}{2}$ → $5 \times 1 =$ ___ — $\frac{4}{5}$ ◯ $\frac{1}{2}$ so round to ___

3. $1\frac{1}{7}$ — $2 \times 1 =$ ___ ← $\frac{1}{7}$ ✕ $\frac{1}{2}$ → $7 \times 1 =$ ___ — $\frac{1}{7}$ ◯ $\frac{1}{2}$ so round to ___

4. $15\frac{9}{20}$ — $2 \times 9 =$ ___ ← $\frac{9}{20}$ ✕ $\frac{1}{2}$ → $20 \times 1 =$ ___ — $\frac{9}{20}$ ◯ $\frac{1}{2}$ so round to ___

5. $921\frac{3}{4}$ — $2 \times 3 =$ ___ ← $\frac{3}{4}$ ✕ $\frac{1}{2}$ → $4 \times 1 =$ ___ — $\frac{3}{4}$ ◯ $\frac{1}{2}$ so round to ___

ESTIMATING ANSWERS

LESSON 3 UNIT 6

You can use rounded numbers to estimate the solutions to problems and equations. When the numbers are large or hard to work with, rounded numbers can help you find an approximate answer.

1742 + 2094 = Round both numbers to nearest hundreds place, then solve.

1700 + 2100 = 3800 so, 1742 + 2094 is **about** 3800

Use the ≈ sign to show an estimate $1742 + 2094 \approx 3800$

You can estimate using any of the four basic operations (addition, subtraction, multiplication, division).

A Estimate these sums.

1. $371 + 563 \approx$ ______

   Round to the nearest ten and add: 370 + 560 = 930

2. $40{,}317 + 62{,}397 \approx$ ______

   Round to the nearest thousand and add: ______ + ______ = ______

3. $8.16 + 6.8 \approx$ ______

   Round to the nearest whole number and add: 8 + 7 = ______

4. $4\frac{2}{3} + 7\frac{7}{8} \approx$ ______

   Round to the nearest whole number and add: ______ + ______ = ______

B Estimate these differences.

1. 973 − 651 ≈ ______

   Round to the nearest hundred and subtract: 1000 − 700 = ______

2. 6,092 − 4, 129 ≈ ______

   Round to the nearest hundred and subtract: ______ − ______ = ______

3. $3\frac{7}{9} - 1\frac{2}{3}$ ≈ ______

   Round to the nearest whole number and subtract: ______ − ______ = ______

4. 12.91 − 6.7 ≈ ______

   Round to the nearest whole number and subtract: ______ − ______ = ______

C Estimate these products.

1. 616 × 199 ≈ ______

   Round to the nearest hundred and multiply: 600 × 200 = ______

2. 47 × 43 ≈ ______

   Round to the nearest ten and multiply: ______ × ______ = ______

3. 4.6 × 2.1 ≈ ______

   Round to the nearest whole number and multiply: ______ × ______ = ______

D Estimate these quotients.

1. 78 ÷ 19 ≈ ______

   Round to the nearest ten and divide: 80 ÷ 20 = ______

2. 960 ÷ 99 ≈ ______

   Round to the nearest hundred and divide: ______ ÷ ______ = ______

3. $3\frac{3}{5} \div \frac{19}{10}$ ≈ ______

   Round to the nearest whole number and divide: ______ ÷ ______ = ______

LESSON 4 UNIT 6

TEST TRACK

A Round to the nearest whole number.

1. $2\frac{5}{8}$ ____________

2. $12\frac{7}{12}$ ____________

3. 6.077 ____________

4. 0.738 ____________

5. 24.705 ____________

6. $3\frac{11}{12}$ ____________

B Round to the nearest hundredth.

1. 174.052 ____________

2. 98.06 ____________

3. 252.775 ____________

4. 1,475.007 ____________

5. 0.3035 ____________

6. 0.999 ____________

C Choose the best estimate.

1. $16.2 \div 3.6 =$

   Ⓐ 4 Ⓑ 2 Ⓒ 5 Ⓓ 6

2. $69.75 \div 34.7 =$

   Ⓐ 25 Ⓑ 2 Ⓒ 2.5 Ⓓ 20

3. $9.125 \div 0.33 =$

   Ⓐ 270 Ⓑ 27 Ⓒ 2.7 Ⓓ 3

4. $0.02 \times 195 =$

   Ⓐ 400 Ⓑ 200 Ⓒ 4 Ⓓ 40

5. $14\frac{7}{8} \div 2\frac{5}{6} =$

   Ⓐ 5 Ⓑ 7 Ⓒ 28 Ⓓ 3

D Circle the greater number.

1. $\frac{5}{6}$ $\frac{7}{8}$
2. $\frac{15}{16}$ $\frac{16}{19}$
3. $3\frac{4}{7}$ $3\frac{5}{8}$
4. $13\frac{2}{3}$ $\frac{40}{3}$
5. $7\frac{3}{7}$ $7\frac{1}{2}$
6. $\frac{5}{12}$ $\frac{6}{13}$

E Solve.

1. Which holds the greater amount, 12 third-cup containers or 7 half-cup containers? Explain your answer.

2. Which holds the lesser amount, 6 fourth-cup containers or a cup-and-a-third? Explain your answer.

LESSON 1 UNIT 7

THINKING ABOUT WORD PROBLEMS

When solving word problems, think about what makes sense. Use these problem solving strategies:

- Look for key words that signal addition, subtraction, multiplication, or division.
- Draw a picture.
- Make a table or chart.
- Solve first using simpler numbers.
- Sort out extra information and ignore it.
- Estimate the solution mentally before solving. Know if you are overestimating or underestimating.
- Check your work. Make sure your answer is reasonable.

A Choose the most reasonable answer.

1. How long is a paper clip?
   A 3 cm.
   B 50 cm.
   C 15 in.
   D 0.5 in.

2. What weighs a ton?
   A a table
   B a deer
   C an elephant
   D a stove

3. How much gas is used driving 100 miles?
   A 5 gallons
   B 50 gallons
   C 5 pints
   D 500 pints

4. What is the height of a doorway?
   A 3 meters
   B 720 inches
   C 7 yards
   D 7 feet

5. How long does it take to walk a mile?
   A 1 hour
   B 20 minutes
   C 3 hours
   D 6 minutes

6. How heavy is a cell phone?
   A 2 lbs.
   B 6 oz.
   C 16 oz.
   D 1.5 lbs.

B **Complete each problem below with numbers that make sense in the situation. Then circle the name of the operation you would use to solve it. You may need more than one operation to solve a problem.**

1. Having grown by almost 10%, the city of Summit has 11,560 more people now than it did five years ago when the population was _____. What is the current population of Summit?

addition **subtraction** **multiplication** **division**

2. Three boys are collecting cans and bottles for recycling. In the past few days, they have collected _____ pounds of cans and bottles. They are paid _____ per pound. How much will they receive?

addition **subtraction** **multiplication** **division**

3. Sara has spent _____ hours in the past week working on an 8-page paper for English class. About how long did she spend on each page?

addition **subtraction** **multiplication** **division**

4. On Thursday, Jane's three highest bowling scores were _____, _____, and _____. What was her average score?

addition **subtraction** **multiplication** **division**

5. Kendrick wants to buy a new jacket that costs _____. He still needs $25.00, but can earn money babysitting at _____ per hour. How long will it take him to earn enough to buy the jacket?

addition **subtraction** **multiplication** **division**

6. Sandy swims at the pool for half an hour each day. He wants to swim 1000 laps to get in shape. If he swims _____ laps per minute, how many days will it take him to accomplish his goal?

addition **subtraction** **multiplication** **division**

LESSON 2 UNIT 7

KEY WORDS AND NUMBERS

When you read a word problem, pay attention to key words and numbers. The words will tell you what to do with the numbers.

Carter has **27** CD's. His friend Patrick has **16** CD's.
How many more CD's does Carter have than Patrick?

The key words "how many more" suggest that you can solve by subtracting.

27 – 16 = 11

Answer: Carter has 11 more CD's than Patrick.

A Look for the key words in these problems, then tell what operation you will use. Solve.

1. Shop-a-Lot Supermarkets are surveying their customers to find out what kind of cola they like best. Out of every five people surveyed, three prefer Cali-Cola and two prefer Fizi-Cola. What percent prefer Cali-Cola?

   Key words: Out of, what percent

   How will I solve? Use division to change a fraction to a percent

   Answer: ____________ prefer Cali-Cola.

2. The Fun Times Amusement Park charges $14 for adults, $10 for children ages 7-18, and $7 for children ages 2-6. Children under 2 are free. The Myers have four children: Tom is 12, Karen is 7, Anne is 5 and David is $1\frac{1}{2}$. How much will they pay in all for the whole family?

   Key words: how much, in all

   How will I solve? ________________________________

   Answer: They will pay ____________ in all.

3. This pie chart shows the kinds of television service used by the families in Jefferson county. What percent of families have no TV? (All the percents should add up to 100).

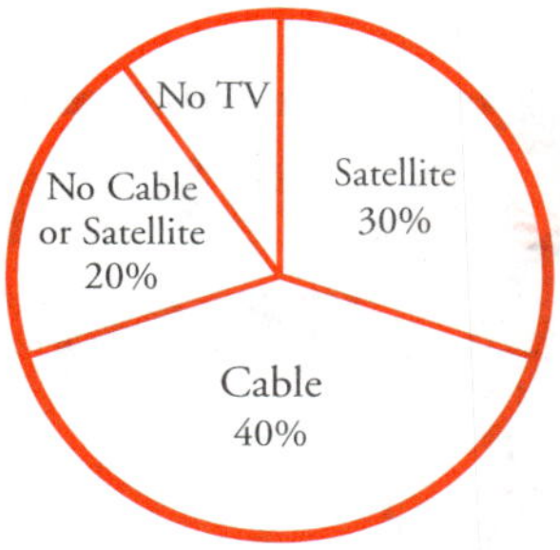

Key words: add up to

How will I solve? ________________________________

Answer: ____________ % have no TV.

B Look for key words and numbers when you solve these problems.

1. Maura spent $76 for hardware. She bought 30 boxes of screws at $1 each. Boxes of nails cost $2. How many boxes of nails did Maura buy?

____________ boxes of nails

2. Two out of three ninth-graders at Valley High School take algebra. There are 81 ninth-graders. How many take algebra?

____________ students take algebra.

3. Michael works in a music store, and is keeping track of CD sales. He finds that a country music CD collection has sold 2,500 copies, while a rock music CD collection has only sold 650 copies. By what number of copies is the country music CD more popular?

____________ copies

LESSON 3 UNIT 7

MORE WORD PROBLEMS

Sometimes you have to do several operations before you find the answer to a word problem.

Pay close attention to the key words and numbers. Write down the information you have, leaving blanks for the information you need to find. Read each problem at least three times.

A **Underline the key words and numbers in each problem. Ask yourself what you need to do to find the answer. Write equations, using blanks for unknown information.**

1. Dan is making pancakes for brunch. He wants to make 2 dozen (24) pancakes, and is using the recipe below. This recipe makes 6 pancakes.

   1 cup pancake mix   1 egg

   $\frac{3}{4}$ cup milk $\frac{1}{2}$ cup blueberries

   How many eggs will he need in all?

   Key words: ____________________

   How will I solve? ____________________

   Equations: ____________________

   Dan will need ________eggs

2. Randy and Barrett work at a farm. Randy can pick 4 pounds of peas in one hour. Barrett can pick 3 pounds in one hour. If they each work six hours a day, how many pounds will they pick working together for one work week (five days)?

   Key words: ____________________

   How will I solve? ____________________

   Equations: ____________________

   In one week, Randy and Barrett will pick ____________ pounds working together.

3. Bob is twice the age of Ellen. Joe is 5 years younger than Bob. If Ellen is 10 years old, how old is Joe?

HINT

Start with these numbers:
Ellen is 10.
Bob is 10 x 2

Key words: ____________________

How will I solve? ____________________

Equations: ____________________

Joe is __________ years old.

B Solve these problems.

1. The variety store is holding a contest to guess how many pennies are in a jar in the window. Susan guesses that the jar is 90% full, and Edmund guesses that it is $\frac{5}{8}$ full. The actual number of pennies is between their two guesses. If the jar can hold 1000 pennies when it's full, which of these numbers could be the actual number in the jar.

A 580 pennies B 600 pennies C 750 pennies D 950 pennies

2. Rachel and Jim are buying trim to install around the windows of their house. They want fluted moldings for the top and two sides of each window, and fluted sills for the bottoms. They have 12 windows measuring 3' wide by 4' high. How many feet of each trim will they need?

__________ feet of moldings __________ feet of sills

3. Carl is mixing paint for a custom paint job. His basic mix calls for twice as much red as yellow, and as much white as red and yellow together. If he uses 4 gallons of red, how much white will he need?

__________ gallons of white paint

LESSON 4 UNIT 7

ESTIMATING ANSWERS TO PROBLEMS

Some problems ask for estimated answers. Look for the words *about, approximately,* and *estimate.* Use the problem solving skills you have learned, but round the numbers before you add, subtract, multiply, or divide.

Jack is studying populations of small towns. He has found that Harborville has **2,781** people and Forestville has **3,468**. **About how many more** people live in Forestville than in Harborville? Round to nearest hundred.

Key words: About how many more

Solve by rounding, then subtracting.

3,468 rounded to nearest hundred: **3,500** 2,781 rounded to nearest hundred: **2,800**

Subtract: 3,500 – 2,800 = 700

About 700 more people live in Forestville than in Harborville.

A Use estimation to answer these problems.

1. Gloria spends about $185 a week for groceries for her family. About how much does she spend in two years (104 weeks)?

Key words: About how much

How will I solve? ______________________________

185 rounded to nearest hundred: __________

104 rounded to nearest hundred: __________

Equation: ______________________

Gloria spends about __________ on groceries in two years.

2. Jason buys a VCR on sale for $97 and a TV on sale for $148. If the regular price of the VCR is $129 and the regular price of the TV is $180, about how much has he saved altogether?

Key words: ____________________

How will I solve? ____

Rounded to nearest ten: 97 _____ 148 _____ 129 ______ 180 _____

Equations: ____

Jason saved about __________.

B Solve these problems. Use estimation.

1. Zelda is shopping for party supplies. She needs enough of each item for 15 people. Use this price list to estimate how much money she should bring with her to the party store.

   Hats: $3.99 for pack of 6

   Plates: $4.25 for pack of 8

   Cups: $3.65 for pack of 4

   Zelda should bring about __________.

2. The standard price for the new SuperZip computer is $1899. The following upgrades are available:

   A. Double RAM $129

   B. Scanner $ 69

   C. CD-rewritable drive $159

   Joshua has $2100 to spend before tax. Which two upgrades can he choose?

   __________ and __________

3. In 92 games, the Rockets baseball team scored 767 runs. Estimate the average number of runs scored per game.

   About __________ runs per game.

LESSON 5 UNIT 7

MAKING JUDGMENTS

When you estimate, pay attention to how close your estimate is likely to be to the exact answer.

Estimate the answer to 947 + 875.

Rounding the numbers to the nearest hundred: 900 + 900 = 1800

Rounding the numbers to the nearest ten: 950 + 880 = 1830

Exact answer: 947 + 875 = 1822.

When the two numbers were rounded to the nearest ten, the estimate was closer to the exact answer.

A Solve these problems. Use estimation.

1. In one day, Burger Heaven serves 168 burgers, 196 orders of fries, and 94 chicken sandwiches. Estimate the total number of items served in one day.

   Rounding numbers to nearest hundred: ____________________

   Rounding numbers to nearest ten: ____________________

   Exact answer: 458 Which estimate is closer? ____________________

   Why?

2. Maria and Robert are making estimates for this equation: 7,624 x 5,396 =
   Maria rounds to the nearest thousand. Robert rounds to the nearest hundred. Whose estimate is likely to be closer to the exact answer? Why?

You can also predict whether your estimate will be an overestimate or an underestimate.

Estimate the answer to 635 + 426.

Round numbers to nearest ten and add: 640 + 430 = 1070

Exact answer: 635 + 426 = 1061

Since both numbers were rounded up, the estimate was higher than actual. It was an overestimate.

B Estimate the answer to each problem. Decide whether it is likely to be an overestimate or an underestimate, and tell why.

1. George receives $183 base salary per week as a waiter. On the average, he takes home another $194 in tips. About how much does he earn in all each week?

   About ______ per week Overestimate or underestimate? ________________

   Why?

2. John has $50 to spend. He wants to buy pants for $39 and a shirt for $17. About how much more money does he need?

   About ______ more Overestimate or underestimate? ________________

   Why?

3. Mel is building a rectangular table 3.9' long and 1.7' wide. What will be the approximate surface area of the table top? Hint: Find area by multiplying length x width.

   About _____ square feet Overestimate or underestimate? ________________

   Why?

LESSON 6 UNIT 7

TEST TRACK

A Choose the best answer to each question.

1. In making a cake, Phil measured $\frac{3}{4}$ cup of sugar to use in the cake and another $1\frac{1}{3}$ cups to use in the frosting. About how much sugar did he use in all?

   Ⓐ $1\frac{1}{2}$ cups Ⓑ 2 cups Ⓒ 3 cups Ⓓ $2\frac{3}{4}$ cups

2. Leah wants to multiply 10,843 by 2,024. Which is the best estimate of the product?

   Ⓐ 11,000 x 2000 Ⓑ 10,500 x 2,500 Ⓒ 10,000 x 2000 Ⓓ 11,000 x 3000

3. Tom needs to estimate the weight of luggage being loaded onto an airplane. There are 145 passengers and 12 crew members. Each has brought an average of 36.7 pounds of luggage onto the plane. What is the best first step in estimating the total weight?

   Ⓐ Round the numbers to the nearest hundred.
   Ⓑ Round the numbers to the nearest ten
   Ⓒ Add the number of passengers and the number of crew.
   Ⓓ Subtract the number of crew from the total number of passengers.

4. To estimate the quotient of 6355 ÷ 83, Teri divided 6400 by 80. Is Teri's estimate greater than or less than the actual?

   Ⓐ Less, because she used a larger divisor and a smaller dividend.
   Ⓑ Less, because she used a smaller divisor and a larger dividend.
   Ⓒ Greater, because she used a larger divisor and a smaller dividend.
   Ⓓ Greater, because she used a smaller divisor and a larger dividend.

5. To estimate the average of 285.2, 174.05, and 690.88, Jim first added 300 + 200 + 700 and then divided by 3. Is Jim's estimate greater than or less than the actual average?

   Ⓐ Less, because the estimated sum is less than actual.
   Ⓑ Less, because the estimated sum is greater than actual.
   Ⓒ Greater, because the estimated sum is less than actual.
   Ⓓ Greater, because the estimated sum is greater than actual.

6. To estimate the product of 1,342 and 204, Sue multiplied 1000 x 200. Is Sue's estimate greater than or less than the actual product?

Ⓐ Less, because she rounded the numbers up.
Ⓑ Less, because she rounded the numbers down.
Ⓒ Greater, because she rounded the numbers up.
Ⓓ Greater, because she rounded the numbers down.

B Answer the question about each problem in your own words.

1. Keith wants to estimate the cost of 3 notebooks that cost $1.89 each plus 8% sales tax. What is a good estimate?

   Explain your answer.

2. What is the volume of a cube that is $2\frac{1}{8}$ inches on each side? What is a good estimate?

   Explain your answer.

REMEMBER
Volume = length x width x height

3. Three shipments weigh 3 tons 1800 pounds, 1 ton 150 pounds, and 4 tons 1500 pounds. What is a good estimate of their total weight?

   Explain your answer.

4. Amy wants to fill her raised garden bed with topsoil. She has ordered 20 cubic feet of topsoil. Will she have enough? Explain how you can use estimation to solve.

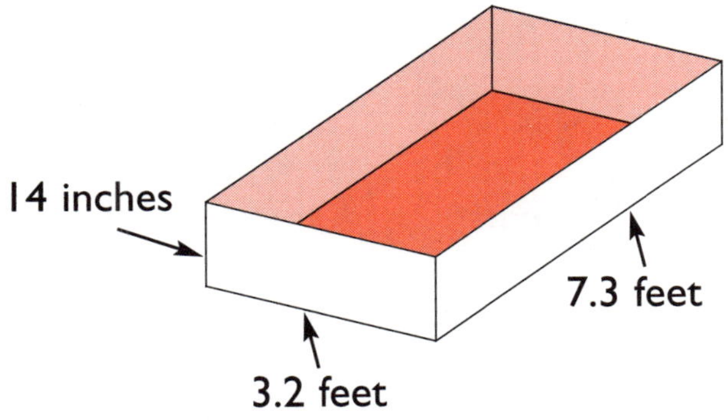

LESSON 1 UNIT 8

UNDERSTANDING PATTERNS

A pattern of numbers is predictable. If you understand how the numbers are related, you can tell what comes next.

2, 4, 6, 8, 10… What is the next number?

In this pattern, you add 2 to each number to get to the next number.

$2 + 2 = 4$ $4 + 2 = 6$ $6 + 2 = 8$ $8 + 2 = 10$

To get to the next number after 10, add 2 again.

$10 + 2 = 12$, so the next number is 12.

Numbers can be related by addition, subtraction, multiplication, or division. When you look at a pattern remember these operations.

A Complete these patterns, using the rule given for each one.

1. add 1 — 3, 4, 5, 6, ________
2. add 2 — 1, 3, 5, 7, ________
3. subtract 3 — 15, 12, 9, 6, ________
4. multiply by 2 — 2, 4, 8, 16, ________
5. subtract 2 — 10, 8, 6, 4, ________
6. divide by 3 — 243, 81, 27, 9, ________

B Complete the pattern. Describe the rule in your own words.

1. 1, 4, 7, 10, ________

2. 20, 16, 12, 8, ________

3. 48, 24, 12, 6, ________

Sometimes two operations are needed to get from one number to the next.

2, 6, 14, 30… What is the next number in this pattern?

In this pattern, you add 1 then multiply by 2 to find the next number.

2 + 1 = 3	3 x 2 = 6
6 + 1 = 7	7 x 2 = 14
14 + 1 = 15	15 x 2 = 30

To find the next number, add 1 then multiply by 2.

30 + 1 = 31 31 x 2 = 62

The next number is 62

C Continue these patterns, using the rule given for each one.

1. 40, 24, 16, 12, ________ Divide by 2, then add 4.

2. 4, 7, 16, 43, ________ Multiply by 3 then subtract 5.

3. 1, 6, 16, 36, ________ Rule: ______________________________

D Use patterns to solve these problems.

1. For his first three-month period on the job, Antonio makes $6.00 per hour. For his second three-month period, he makes $6.25 per hour. What will he make per hour for his <u>fourth</u> three-month period? Pattern rule: add 25 cents.

   $ ________ per hour

2. Althea is picking strawberries. She picks handfuls of 5 berries to put in her basket, but out of each handful she eats 2. After she has picked 6 handfuls of 5 berries each, how many berries are in her basket? Pattern rule: add 5 then subtract 2.

   ________ berries

LESSON 2 UNIT 8

FINDING RULES FOR PATTERNS

Every pattern follows a rule.

10, 20, 30, 40, 50,....... **Rule:** Add 10 to each number to get to the next number.

2, 4, 8, 16, 32,..... **Rule:** Multiply each number by 2 to get to the next number.

To figure out how a pattern works:

- **Ask** yourself how the first two numbers are related.
- **Think** of possible rules to explain how the first two numbers are related (try addition, subtraction, multiplication, division).
- **Test** the rule or rules for all the numbers in the pattern.
- **Decide** which rule works for all the numbers.
- **Apply** the rule to find the next number.

5, 10, 15, 20, 25,...... What is the next number in this pattern?

- First two numbers are 5 and 10. $5 + 5 = 10$, $5 \times 2 = 10$.

- Possible rules: Add 5 to each number, OR multiply each number by 2.
- Add 5: $10 + 5 = 15$ Multiply by 2: $10 \times 2 = 20$

  $15 + 5 = 20$ $20 \times 2 = 40$

  $20 + 5 = 25$ $40 \times 2 = 80$

- Only one rule works for all numbers: Add 5
- $25 + 5 = 30$ The next number in the pattern is 30.

A Find the rule for each pattern.

1. 2, 6, 18, 54,....

- How are 2 and 6 related? $2 + 4 = 6$ $2 \times 3 = 6$
- Possible Rules: Add 4 Multiply by 3
- Test rules: ______________________________
  ____
  ____
- Which rule works? ______________________________
- Apply rule: ______________________________
- Next number in pattern: 162

2. 40, 20, 10, 5,.....

- How are 40 and 20 related? $40 - 20 = 20$ $40 \div 2 = 20$
- Possible rules: Subtract 20 Divide by 2
- Test rules: ______________________________
  ____
  ____
- Which rule works? ______________________________
- Apply rule: ______________________________
- Next number in pattern: $2\frac{1}{2}$

LESSON 3 UNIT 8

MATHEMATICAL REASONING

The rules for patterns can be written in math language.

1, 3, 5, 7, 9,.... **Rule:** Add 2 **Rule in math language:** $n + 2$

n is a **variable**, and stands for each number in the pattern.

When you use the rule, put each number in the place of *n*, then solve the equation you have made.

$n + 2 =$

A Continue the pattern using $n + 2$

1. $5 + 2 = 7$; $7 + 2 =$ ______; $9 + 2 =$ ______;

$11 + 2 =$ ______; ______ $+ 2 =$ ______; ______ $+ 2 =$ ______;

______; ______; ______

Variables can also be used any time there is an unknown number.

$2 + 3 = x$	Since you know that $2 + 3 = 5$, you can solve for *x*.
$2 + 3 = 5$	
$5 = x$	so x must equal 5

$x = 6 - 2$	Since you know that $6 - 2 = 4$, you can solve for x.
$6 - 2 = 4$	
$x = 4$	so x must equal 4

B Solve these equations for x.

1. $x = 7 + 4$ $7 + 4 =$ ______ $x =$ ______

2. $x = 8 - 5$ $8 - 5 =$ ______ $x =$ ______

3. $x = 4 \times 2$ $4 \times 2 =$ ______ $x =$ ______

4. $x = 6 \div 2$ $6 \div 2 =$ ______ $x =$ ______

Sometimes the variable is on the same side of the equation as other numbers. To get x by itself, subtract 4 from both sides.

$4 + x = 6$

$4 + x - 4 = 6 - 4$

$x = 2$

REMEMBER

You can add, subtract, multiply, or divide both sides of an equation in the same way and keep the sides equivalent.

C Solve these equations for the variable.

1. $x + 8 = 10$ To get x by itself, subtract 8 from both sides

   $x + 8 - 8 = 10 - 8$

   $x =$ ______

2. $y - 3 = 1$ To get y by itself, add 3 to both sides.

   $y - 3 + 3 = 1 + 3$

   $y =$ ______

3. $z \times 2 = 10$ To get z by itself, divide both sides by 2.

   $z \times 2 \div 2 = 10 \div 2$

   $z =$ ______

4. $a \div 4 = 3$ To get a by itself, multiply both sides by 4.

   $a \div 4 \times 4 = 3 \times 4$

   $a =$ ______

LESSON 4 UNIT 8

MORE THAN ONE VARIABLE

Some equations have more than one variable.
Barry has three more books than Don. How many books does Barry have?

***b* = the number of books Barry has** ***d* = the number of books Don has**

Write the problem in equation form: $b = d + 3$

If Don has 4 books, how many books does Barry have?

Put 4 where you have *d* in the equation: $b = 4 + 3$

Then solve for *b*: $b = 7$

This table shows some possible values for *d*.
Use the equation $b = d + 3$
to fill in the empty boxes.

d	*b*
4	7
2	
8	

$2 + 3 = 5$, so if $d = 2$, $b = 5$

$8 + 3 = 11$, so if $d = 8$, $b = 11$

Notice that each time one variable changes, the other variable changes too.

A Fill in these tables. Use the equations given.

1. $a = b + 5$

b	*a*
3	8
4	
1	

2. $m = n \times 3$

n	*m*
4	12
6	
5	

3. $x = y - 2$

y	*x*
5	3
9	
3	

4. $j = k \div 2$

k	*j*
8	4
6	
10	

B Solve these problems using the equations and tables given.

1. Monica has three times as many baseballs as Belinda. If Belinda has two baseballs, how many does Monica have? If Belinda has four baseballs, how many does Monica have? If Belinda has six baseballs, how many does Monica have?

$m = b \times 3$

b	m
2	6
4	☐
6	☐

If Belinda has 2, Monica has ________.

If Belinda has 4, Monica has ________.

If Belinda has 6, Monica has ________.

2. Every day Cara and Dan bring pencils to school. Dan always brings twice as many pencils as Cara. Fill in the missing values in this table.

$d = 2 \times c$

d	c
☐	4
10	☐
☐	3

3. Paul and Annika's ages add up to 58. If Paul is 27, how old is Annika?

$p + a = 58$

$a =$ ________

Annika's age is ________.

HINT

Subtract p from both sides to isolate a.

LESSON 5 UNIT 8

TEST TRACK

A Continue the patterns using the formulas given.

1. $n + 4$ — 1, 5, 9, ____, ____, ____

2. $2n + 4$ — 1, 6, ____, ____, ____

3. $2n - 1$ — 2, ____, ____, ____

4. $n + 7$ — 1, 8, ____, ____, ____

5. $x + \frac{1}{2}$ — $\frac{1}{2}$, 1, ____, ____, ____

B Solve for x. Show your work.

1. $2x = 27.6$

2. $4x = 30$

3. $4x + 2 = 26$

4. $2x - 3 = 13$

5. $\frac{1}{x} + \frac{2}{3} = 1$

C Answer the question about each problem in your own words.

1. Abby has decided to exercise every day. On her first day, she plans to exercise for 5 minutes. She will then add 5 minutes to her exercise schedule each day. On which day will she reach a half hour? On which day will she reach one hour? Make a chart, draw a picture, or write an equation to explain your answer.

2. Jeff is trying to run farther and farther each day. He started with a 1/2 mile run yesterday. Today, he ran a quarter-mile less than twice as far. If he repeats his performance every day by running a quarter mile less than twice as far as the previous day, how far will he run on the day after tomorrow? Make a chart, draw a picture, or write an equation to explain your answer.

3. Frank is a waiter. He gets to keep 75% of his tips. He shares the rest with the dishwashers. On Monday night his total tips amounted to $8. If his total tips double every night this week, how much will the dishwashers get to share on Thursday night? Make a chart, draw a picture, or write an equation to explain your answer.

PRACTICE RUN

SKILL: NUMBER WORDS, PLACE VALUE

Choose the correct answer to each question.

1. What is the numeral for twelve hundred two?

Ⓐ 11,202 Ⓒ 12,202
Ⓑ 1,220 Ⓓ 1,202

2. What is the numeral for nine million nine hundred?

Ⓐ 9,009,000 Ⓒ 9,000,900
Ⓑ 900,900 Ⓓ 909,000

3. What is the numeral for two million twenty thousand two?

Ⓐ 2,020,002 Ⓒ 2,000,202
Ⓑ 2,200,002 Ⓓ 2,002,002

4. What is the numeral for six hundred thousand five hundred fifty-six?

Ⓐ 6,005,056 Ⓒ 600,556
Ⓑ 6,000,556 Ⓓ 605,560

5. What is the numeral for seventy-seven thousand forty?

Ⓐ 70,740 Ⓒ 77,400
Ⓑ 77,040 Ⓓ 70,704

6. What is the numeral for 1×10^6?

Ⓐ 10,000 Ⓒ 1,000,000
Ⓑ 100,000 Ⓓ 10,000,000

7. What is the numeral for $(2 \times 10^3) + (1 \times 10^1)$?

Ⓐ 2,010 Ⓒ 20,010
Ⓑ 20,100 Ⓓ 2,001

8. What are the words for 3, 044, 003?

Ⓐ thirty million, forty-four thousand, three
Ⓑ three hundred forty-four thousand, three
Ⓒ three million, forty-four thousand, three
Ⓓ three hundred four thousand, four hundred three

9. What are the words for $(5 \times 10^5) + (5 \times 10^2) + (5 \times 1)$?

Ⓐ fifty thousand, five hundred, five
Ⓑ five hundred thousand, five thousand, five
Ⓒ five million, five hundred, five
Ⓓ five hundred thousand, five hundred, five

10. What does the 8 in 7, 087,005 represent?

Ⓐ 8×10^2
Ⓑ 8×10^3
Ⓒ 8×10^4
Ⓓ 8×10^5

SKILL: MULTIPLES, FACTORS, AND PRIMES

Choose the correct answer to each question.

1. What are the factors of 14?

   Ⓐ 1, 2, 7, 14 Ⓒ 1, 6, 8, 14
   Ⓑ 1, 4, 7, 14 Ⓓ 1, 14

2. What are the common factors of 39 and 54?

   Ⓐ 1,3 Ⓒ 1,3,13
   Ⓑ 1,3,7 Ⓓ 1,13,18

3. What is the Greatest Common Factor of 36 and 63?

   Ⓐ 3 Ⓒ 9
   Ⓑ 6 Ⓓ 12

4. What is the Least Common Multiple of 33 and 66?

   Ⓐ 3 Ⓒ 11
   Ⓑ 66 Ⓓ 22

5. Which factor of 72 is missing from this set? {1, 2, 3, 4, 6, 8, 12,18, 24, 36, 72}

   Ⓐ 5 Ⓒ 10
   Ⓑ 9 Ⓓ 14

6. Which is a prime number?

   Ⓐ 12 Ⓒ 31
   Ⓑ 21 Ⓓ 39

7. Which is not a prime number?

   Ⓐ 2 Ⓒ 4
   Ⓑ 3 Ⓓ 5

8. Which is a set of prime numbers?

   Ⓐ {11, 13, 17}
   Ⓑ {10, 11, 13}
   Ⓒ {11, 14, 17}
   Ⓓ {17, 19, 22}

9. What is the prime factorization of 80?

   Ⓐ $2 \times 2 \times 2 \times 10$
   Ⓑ $2^3 \times 10$
   Ⓒ $2^4 \times 5$
   Ⓓ $2^4 \times 5^2$

10. What is the Prime Factorization of 72?

   Ⓐ 8×9
   Ⓑ $6^2 \times 2^2$
   Ⓒ $2 \times 2 \times 3 \times 3 \times 3$
   Ⓓ $2^3 \times 3^2$

SKILL: ROUNDING

Choose the correct answer to each question.

1. What is 52.6 rounded to the nearest whole number?

Ⓐ 52
Ⓑ 53
Ⓒ 52.5
Ⓓ 56

2. What is 245.09 rounded to the nearest ten?

Ⓐ 245
Ⓑ 246
Ⓒ 249
Ⓓ 250

3. What is 582.63 rounded to the nearest hundred?

Ⓐ 582.60
Ⓑ 500
Ⓒ 583.60
Ⓓ 600

4. What is 456,730 rounded to the nearest ten thousand?

Ⓐ 400,000
Ⓑ 450,000
Ⓒ 460,000
Ⓓ 500,000

5. What is 5 7/8 rounded to the nearest whole number?

Ⓐ 5
Ⓑ 6
Ⓒ 7
Ⓓ 8

6. What is 89,640 rounded to the nearest thousand?

Ⓐ 80,000
Ⓑ 89,000
Ⓒ 90,000
Ⓓ 90,600

7. What is 12.65 rounded to the nearest tenth?

Ⓐ 10.0
Ⓑ 12.5
Ⓒ 12.6
Ⓓ 12.7

8. What is 5,038 rounded to the nearest ten?

Ⓐ 5,000
Ⓑ 5,030
Ⓒ 5,040
Ⓓ 5,300

9. What is 49.7 rounded to the nearest whole number?

Ⓐ 47
Ⓑ 49
Ⓒ 50
Ⓓ 51

10. What is 6,450,987 rounded to the nearest hundred thousand?

Ⓐ 6,500,000
Ⓑ 6,000,000
Ⓒ 6,400,000
Ⓓ 6,450,000

SKILL: EQUIVALENT FRACTIONS

Choose the correct answer to each question.

1. Which fraction is equivalent to 4/7?

   Ⓐ 20/35 Ⓒ 7/4
   Ⓑ 12/20 Ⓓ 42/72

2. Which fraction is equivalent to 9/15?

   Ⓐ 3/5 Ⓒ 18/24
   Ⓑ 1/3 Ⓓ 3/3

3. Which fraction is NOT equivalent to 3/4?

   Ⓐ 12/16 Ⓒ 75/100
   Ⓑ 15/24 Ⓓ 6/8

4. Which fraction is NOT equivalent to 5/13?

   Ⓐ 15/39 Ⓒ 10/25
   Ⓑ 50/130 Ⓓ 20/52

5. What is another way to express 6/8?

   Ⓐ three halves Ⓒ eight sixths
   Ⓑ two out of four Ⓓ three out of four

6. What is another way to express 14/28?

   Ⓐ one half
   Ⓑ four out of seven
   Ⓒ three halves
   Ⓓ three out of four

7. Which pair of fractions is NOT equivalent?

   Ⓐ 5/6, 30/36 Ⓒ 4/5, 12/20
   Ⓑ 7/8, 14/16 Ⓓ 5/14, 10/28

8. Which pair of fractions is NOT equivalent?

   Ⓐ 4/21, 12/63
   Ⓑ 6/9, 18/28
   Ⓒ 8/11, 24/33
   Ⓓ 6/15, 30/75

9. Which is a common factor of both the numerator and denominator in 34/51?

   Ⓐ 17
   Ⓑ 13
   Ⓒ 12
   Ⓓ 11

10. Which is a common factor of both the numerator and denominator in 16/40?

    Ⓐ 5
    Ⓑ 8
    Ⓒ 10
    Ⓓ 16

GLOSSARY

A

addend - A number that is added to another number to find a sum is called an addend. In 4 + 6 = 10, 4 and 6 are addends; 10 is the sum.

C

common factors - Factors are numbers that are multiplied together to find a product. Common factors are those factors which two products have in common. Common factors for 12 and 18 are 2, 3, and 6.

composite number - Any number that has factors other than 1 and itself are composite. A number that has only 1 and itself as factors is prime. 4 is a composite number because its factors are 1, 2, and 4. 5 is a prime number because its factors are 1 anad 5.

cubed - Any number with the exponent 3 is said to be cubed. 2^3 is read "two cubed." The cube of a number is the product of the number used 3 times as a factor. For example, the cube of 2 is 8 because 2 x 2 x 2 = 8.

D

decimal number - A decimal is a fraction with a denominator of 10 or a power of 10 expressed using a decimal point instead of a denominator. For example, 7/10 = .7, 34/100 = .34, and 3/1000 = .003 .

denominator - The number that is below or to the right of the line in a fraction is called the denominator. In 2/7, 7 is the denominator.

digit - any of the counting numbers from 0 to 9

difference - The amount that is left after one number is subtracted from another is called the difference. In 11-6 = 5, 5 is the difference.

dividend - A number that is divided by another number is called the dividend. In 21 ÷ 7 = 3, 21 is the dividend.

divisor - A number by which another number is divided. In 36 ÷ 9 = 4, 9 is the divisor.

E

equation - A number sentence that shows that two amounts or figures are equal is called an equation. 4 - 4 = 0, 2x + 3 = 9, and $a^2 + b^2 = c^2$ are examples of equations.

equivalent fractions - Two different fractions that have the same value are said to be equivalent fractions. 1/2 and 2/4 are equivalent fractions.

estimating - Making a careful, but inexact, guess about the value or cost of something is called estimating. Estimating is used to check the reasonableness of numerical solutions to word problems.

expanded form - A number written such that each digit in the number is expressed as a multiple of 10 is said to be in expanded form. 365 is in standard form. 300 + 60 + 5, and $(3 \times 10^2) + 6 \times 10^1)$ $+ (5 \times 10^0)$ are in expanded form.

exponent - A power associated with a base that indicates how many times the base is used as factor. In 3^4, 3 is the base, and 4 is the exponent, or power. $3^4 = 3 \times 3 \times 3 \times 3 = 81$

F

factor - A number that is multiplied by another number or numbers to find a product is called a factor. 3, 4, and 6 are factors in $3 \times 4 \times 6 = 72$.

formula - A group of symbols or figures that show the elements of something is called a formula. The formula for finding the area of a rectangle is $A = l \times w$, where A stands for area, l stands for length, and w stands for width.

G

GCF - see Greatest Common Factor

Greatest Common Factor - Given the shared factors of two or more products, the one that is greatest in value is the Greatest Common Factor. For example, the factors of 14 are 1, 2, 7 and 14. The factors of 21 are 1, 3, 7, and 21. The Greatest Common Factor of 14 and 21 is therefore 7.

I

inequality - A number sentence comparing two things which are different in value is said to be an inequality. $7 > 3$, and $2x + 4 > 10$ are inequalities.

integer - The whole numbers and their negatives are called integers. . . . -3, -2, -1, 0, 1, 2 . . . are integers.

L

LCD - see Least Common Denominator

LCM - see Least Common Multiple

Least Common Denominator - Given the shared multiples of two or more denominators, the one that is least in value is the Least Common Denominator. For example, the multiples of 3 are 3, 6, 9, 12, 15 . . .. Multiples of 5 are 5, 10, 15 . . .. Therefore, the least common denominator of 1/3 and 2/5 is 15.

Least Common Multiple - Given the shared multiiples of two or more numbers, the one that is least in value is the Least Common Multiple. For example, the multiples of 4 are 4, 8, 12, 16, 20, 24 . . .. Multiples of 6 are 6, 12, 18, 24 . . .. Therefore, the least common multiple of 4 and 6 is 12.

like fractions - Fractions which have the same denominator are said to be like fractions. 1/6 and 5/6 are like fractions.

M

minuend - The number from which another number is subtracted is called the minuend. In $7 - 3 = 4$, 7 is the minuend.

multiple - A number that contains another number an exact number of times with no remainder is a multiple of that number. Multiples of 5 are 5, 10, 15, 20, and so on.

N

numerator - The number that is above or to the left of the line in a fraction is called the numerator. In 2/7, 2 is the numerator.

P

percent - A hundredth of a number is a percent. The symbol for percent is %. 8/100 is 8%.

prime factorization - Finding all of the prime factors of a number is called prime factorization. The prime factorization of 24 is 2 x 2 x 2 x 3, or 2^3 x 3.

prime number - A number that has only 1 and itself as factors is called a prime number. 2, 3, and 5 are examples of prime numbers.

product - The amount that results from multiplying two or more numbers together is called the product. In 2 x 3 = 6, 6 is the product. 2 and 3 are factors.

quotient - The amount that results from dividing one number by another is called the quotient. In 14 ÷ 7 = 2, 2 is the quotient. 14 is the dividend and 7 is the divisor.

R

rational number - Fractions that can be written with integers as the numerator and denominator are rational numbers. 1/4 and 5/16 are examples of rational numbers.

S

squared - Any number with the exponent 2 is said to be squared. 3^2 is read "three squared." The square of a number is the product of the number used 2 times as a factor. For example, the square of 3 is 9 because 3 x 3 = 9.

standard form - see expanded form

subtrahend - A number subtracted from another number is called the subtrahend. In 6 - 4 = 2, 4 is the subtrahend. 6 is the minuend and 2 is the difference.

sum - The result of adding two or more addends is called their sum. In 2 + 4 + 5 = 11, 11 is the sum. 2, 4, and 5 are addends.

U

unlike fractions - Fractions which have different denominators are said to be unlike fractions. 1/7 and 5/6 are unlike fractions.

variable - A symbol in a numerical expression that represents a value that can change. In 2x + y = 100, x and y are variables.

W

whole number - Any number zero and greater containing no fractions or decimals is called a whole number. 0, 1, 2, 3, 4 . . . are whole numbers.